Electric Wiring:
Domestic

Electric Wiring: Domestic

Tenth Edition

A. J. Coker, W. Turner

revised by
B. Scaddan

Newnes
An imprint of Butterworth-Heinemann Ltd
Linacre House, Jordan Hill, Oxford OX2 8DP

A member of the Reed Elsevier group

OXFORD   LONDON   BOSTON
MUNICH   NEW DELHI   SINGAPORE   SYDNEY
TOKYO   TORONTO   WELLINGTON

First published 1940
Second edition 1944
Third edition 1947
Fourth edition 1954
Fifth edition 1957
Sixth edition 1967
Seventh edition 1969
Reprinted 1975, 1976, 1979
Eighth edition 1983
Reprinted with revisions 1985
Ninth edition 1989
Reprinted 1989, 1991, 1992
Tenth edition 1992
Reprinted 1993

British Library Cataloguing in Publication Data
Electric Wiring: Domestic. – 10 Rev. ed
  I. Coker, A. J.   II. Turner, W.
  III. Scaddan, Brian
  621.31924

ISBN 0 7506 0804 8

Printed and bound in Great Britain

Preface

This is the tenth edition of a book which has for many years been in constant demand as a clear and reliable guide to the practical aspects of domestic electric wiring. Intended for electrical contractors, installation engineers, wiremen and students, its aim is to provide essential up to date information on modern methods and materials in a simple, clear and concise manner.

The main changes in this edition are those necessary to bring the work into line with the 16th Edition of the *Regulations for Electrical Installations* issued by the Institution of Electrical Engineers.

Chapters 1 to 3 introduce the basic features of domestic installations; explain power and current ratings, cable and accessory sizes, and circuit protection; and the fitting of switches, fuses, circuit-breakers, etc. Chapters 4 to 6 deal in detail with the main types of domestic wiring work, covering lighting, power, socket-outlets, and the connection of appliances. Fluorescent lighting and 'off-peak' electric heating systems are also covered.

Chapters 7 to 9 deal with the principal wiring systems available for domestic use, including steel and PVC conduit, PVC cable, and the mineral-insulated copper-sheathed system.

Chapter 10 describes the earthing requirements and the protective multiple earthing (PME) system which is now being more widely applied; it also discusses earth-leakage circuit-breakers, which are becoming increasingly common. Chapter 11 explains the inspection and tests required on completed installations, including the earth-fault loop-impedance and ring-circuit continuity tests which are now covered in greater detail in the Regulations.

Acknowledgements

The publisher's thanks are due to the following manufacturers who have kindly supplied illustrations and up to date information on their products:

Ashley Accessories Ltd.
Barton Conduits Ltd.
British Aluminium Co. Ltd.
British Insulated Callender's Cables Ltd.
Crabtree Electrical Industries Ltd.
Creda Ltd.
Dimplex Heating Ltd.
Dorman Smith Ltd.
Edgecumbe Ltd.
Egatube Ltd.
Evershed & Vignoles Ltd.
Findlay, Durham & Brodie Ltd.
GEC Ltd.
Gilflex Conduits Ltd.
Maclaren Controls Ltd.

Midland Electric Manufacturing Co. Ltd.
MK Electric Ltd.
Nettle Accessories Ltd.
Parkinson Cowan Heating Ltd.
Pirelli Cables Ltd.
Pyrotenax Ltd.
Santon Ltd.
Satchwell Appliance Controls Ltd.
Geo. H. Scholes & Co. Ltd. (Wylex)
Simplex—GE Ltd.
Tenby Electrical Accessories Ltd.
Unidaire Engineering Ltd.
Volex Electrical Products Ltd.
Walsall Conduits Ltd.

Contents

1   Wiring circuits simply explained   1

2   Ratings, cable sizes and circuit protection   12

3   Domestic installation practice   28

4   Wiring lighting points   39

5   Wiring socket-outlets and portable appliances   60

6   Wiring fixed appliances   73

7   Survey of modern wiring systems   106

8   Installation of conduit systems   118

9   Installation of PVC-sheathed and MICS cable   135

10   Safe and efficient earthing   148

11   Inspection and testing   164

    Index   180

1 Wiring circuits simply explained

An electrical circuit is the whole path along which an electric current may flow. This path may be divided into three parts:

1   The source of electricity (e.g. generator, transformer or battery), which supplies energy to the supply terminals of the circuit.
2   The consuming devices, which convert the electrical energy into the desired form, such as heat or light.
3   The wires and control gear, which convey the electricity from the supply terminals on the premises to the consuming devices.

The simplest circuit contains three wires, one connecting the neutral side of the supply to the apparatus, one connecting the apparatus to a single-pole switch, and one connecting the switch to the supply (Fig. 1.1).

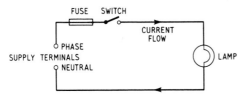

Fig. 1.1. The simplest lighting circuit consisting of leads, fuse, switch and lamp connected to the supply terminals. The single-pole switch is connected in the phase wire

In certain cases, a double-pole switch is used to isolate both wires of the supply from the apparatus, and then a four-wire circuit is needed.

Parallel circuits

Circuits can be built up from this simple three-wire circuit. For instance, if two lighting points are to be controlled from one single-pole switch, a loop wire is taken from each terminal of the first point and connected to the corresponding terminal of the second point. When the switch is closed the current has two paths, one through each lamp, so that both are energised. The two lamps are then said to be connected in parallel (Fig. 1.2).

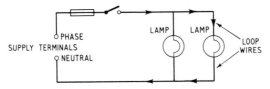

Fig. 1.2. In the simplest parallel circuit both lamps have the full voltage across them

When several circuits have to be supplied from the same mains, the switch wires are all connected to one side of the supply and the return (or neutral) wires from the apparatus to the other side. These circuits are then in parallel.

The full electrical pressure of the supply (in volts) is applied to each circuit and the amount of power (in watts) taken by any one circuit is independent of the power taken by the other circuits.

Colour coding of wires

In order to distinguish wires from each other, the insulation is normally coloured. On single-phase systems the phase is coloured red, the neutral black and the earth green-and-yellow.

With polyphase systems (i.e. where there is more than one phase supply terminal), yellow and blue are also used to denote phase wires. In certain types of wiring the earth conductor (now called the protective conductor) is simply a bare wire, or strands of wire, and this is not colour-coded except at the terminations (see Chapter 5).

The red (phase) wire is the one which is normally broken by the switch; this ensures that the apparatus is electrically disconnected from the supply when the switch is open

Colour coding of flexible cables

The above colours refer to fixed wiring with non-flexible conductors. For flexible cables the colours are brown, blue and green-and-yellow for the phase, neutral and protective (earth) conductors respectively.

It should be noted that the term 'live' normally includes both the phase and neutral conductors, but it does not include the neutral on a protective multiple earthing (PME) system. IEE Regulations Part 2 refers. The PME system is described in Chapter 10.

Protection of circuits and apparatus

It is most important to make sure that any fixed installation or portable apparatus is properly protected against overloading, and that there is no risk of fire in the event of a fault developing.

Fuses

One common method of protecting circuits and apparatus is by inserting a fuse in the phase conductor. A fuse is a device for protecting a circuit against damage due to excessive current flowing in it, operating by opening the circuit on the melting of the fuse element. The two main types of fuse are known as the semi-enclosed (or rewirable) and cartridge types.

A *semi-enclosed* fuse has a fuse holder or link of an incombustible material such as porcelain or moulded resin, and a fine wire between the two contacts, partly enclosed in the fuse-holder or in an asbestos tube. The capacity of the fuse is altered by using a fuse wire of different gauge.

A *cartridge* fuse has a similar fuse-holder or link, but the fuse element is contained in a cartridge of incombustible material filled with fine, arc-suppressing sand or other suitable material. Cartridges of various current rating are available.

Fuses are usually installed in a fuseboard or consumer unit, with the main isolating switch, adjacent to the meter.

Portable apparatus may also be protected by using the standard 13 A rectangular-pin plug, which incorporates a cartridge fuse in the phase side. These cartridge fuses are available in 2, 3, 5, 10 and 13 A capacities. They are colour-coded as well as labelled (2, 5, 10 A are black, 3 A are red and 13 A brown).

Miniature circuit-breakers

As an alternative to fuses, miniature circuit-breakers (MCBs) may be used to protect circuits from excess current. These are automatic switches which open when the current flowing through them exceeds the value for which they are set. Variation of rating can be obtained by making the operating mechanism (which is usually an electromagnet and a bi-metal strip) operate at different currents. MCBs may be installed in a distribution board or consumer unit in a similar manner to fuses, and some boards are, in fact, designed to accommodate either.

Earthing

A fault or accidental damage may cause live conductors to come into contact with the metal casing of apparatus or other accessible metal parts of an installation.

It is, therefore, necessary to ensure that should this happen the fault current does not flow for long enough to cause damage or fire, and that there is no risk of anybody receiving a shock should they be touching the metalwork. This is achieved by connecting any such metalwork to the general mass of earth, so that the resistance of the path to earth is low enough to ensure that sufficient fault current passes to operate the protective device and also that the fault current takes this earth path rather than through a person touching the metalwork (see Chapter 10 for details).

Distribution fuseboards

One obvious way of connecting up circuits would be to run a pair of conductors round the building and to tap off current where it is required.

However, this should on no account be done. The supply ends of the various lighting and power circuits should be brought back to a convenient point in the building and there connected to a distribution fuseboard, each individual circuit protected by a fuse or circuit-breaker.

An installation in a normal house or flat will usually require only one distribution fuseboard or consumer unit (except when the electricity tariff calls for two or more separate meters). The supply is controlled by a consumer's main double-pole switch, which is usually included in the consumer unit, so that the whole installation can be made 'dead' when required. A pair of main cables, sufficiently large to carry the maximum current taken by the installation are connected from the electricity board's service fuse and neutral link through their meter to the consumer's main switch.

Large installations may require several distribution fuseboards, each supplying one floor or section of the premises. In such cases, the incoming mains are taken to a main distribution board, where they connect to a number of large fuses or circuit breakers protecting the outgoing circuits. Sub-main cables connect the main board to smaller branch distribution boards, which contain smaller fuses (or MCBs), protecting the actual lighting and power final circuits. Such cases are not very common on domestic installations.

A.C. mains supply

Power supplies are generated almost everywhere as alternating current (a.c.), which means that the current is changing direction continually. In the U.K., this change of direction occurs 50 times per second. The machines which generate the power are known as alternators and they have three identical sets of windings in which the current is generated. One end of each winding is connected to a common, or star, point which is known as the neutral. The other ends of the windings are brought out to the three wires or phases of the supply cables. For identification purposes, these phases are colour-coded red, yellow and blue and the currents which are transmitted in each phase have a phase displacement of 120 degrees.

Power supplies to towns and villages are provided from the power stations all over the country, via a system of overhead and underground mains and transformers which reduce the voltage in steps from the transmission voltage to the normal mains voltage at the consumers.

The National Grid Company is responsible for the transmission of power in bulk, at 400 kV, on the Super Grid System. The local Electricity Boards are responsible for the distribution of power from the grid stations to domestic and industrial consumers. This is done by using a network of overhead lines and underground cables (usually at 132 kV and 33 kV) to take the power to the main load centres, and 11 kV overhead lines and underground cables to distribute the power to individual load centres, where secondary sub-stations reduce the voltage to mains potential. From these sub-stations (which can consist of pole-mounted transformers and isolating gear or ground-mounted transformers and switch-gear), low voltage overhead and underground mains are taken to the consumer's supply terminals (Fig. 1.3).

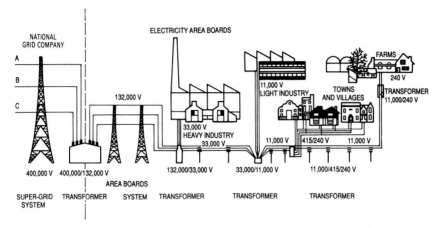

Fig. 1.3. Distribution of electricity

A.C. power is distributed mainly on three-phase networks, although in some rural areas single-phase supplies only are available. Thus, on the low-voltage side of most local transformers we have four terminals: red, yellow and blue phases and neutral. Between each phase and the neutral there is a voltage of 240 V (the standard voltage in Britain); between the phases the voltage is 415 V. For this reason, the secondary output voltage of transformers is given as 415/240 V (Fig. 1.4).

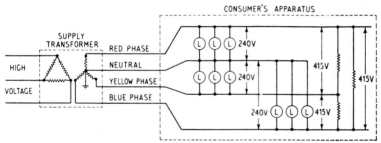

Fig. 1.4. Three-phase, four-wire system of a.c. supply. The phases are star connected, with the neutral point earthed. The voltage between any phase and neutral is 240 V, while the voltage between any two phases is 415 V

The supply authorities are required to maintain the voltage at the consumer's supply terminals within plus or minus 6 per cent of the nominal voltage, i.e. a 240 V single-phase supply must be kept between 224 and 256 V.

The various systems for distributing power to consumers are defined in Part 2 of the IEE Regulations and in the UK are designated TT, TN-S or TN-C-S.

Most domestic consumers are provided with a single-phase supply, unless exceptionally heavy loading is anticipated. Many industrial premises have a three-phase supply since the load may be high and most heavy machinery is powered by three-phase motors.

Power transformers are more efficient if the load on each phase is approximately the same and, therefore, single-phase services are normally connected to alternate phases and three-phase consumers are encouraged to balance their loads over the three phases.

The local boards provide underground-cable or overhead-line services which terminate at a convenient point within the premises. Overhead services are terminated on a bracket high up on a wall of the property and insulated leads taken through the wall to the meter position. Underground services are usually brought through below floor level via a duct. The overhead or underground service leads are taken into the electricity board's main fuse. This is usually of 100 A rating for domestic properties. From the fuse the supply is taken to the meter. Where three-phase supplies are provided, three main fuses are used with one composite meter.

With underground services, either the lead sheath of the cable or, as in the case of plastic cables, the wire armour is used to provide an earth. A separate wire is usually bound and soldered to the sheath or armour at the terminal position and taken to an earth-connector block. All protective conductors in the property are brought back to this block.

With overhead services, an earth block may be provided if PME is adopted (see Chapter 10 for details), or else an earth-electrode, in conjunction with a residual current circuit-breaker, must be used. In some cases a separate overhead earth conductor is provided.

In large blocks of flats or offices, the services to each floor are provided by 'rising mains'. For this purpose, ordinary paper or PVC-insulated mains cable can be used, or mineral insulated cable or bare copper busbars in protected trunking. Sub-services are teed off at the various floor levels (see Chapter 9).

Consumers' meters are often arranged so that they can be read without entering the premises. In houses or small blocks of flats this is usually achieved by having the meter on an outside wall and visible through a small vandal-proof window. In multi-story blocks of flats the meters are often fitted in the riser duct running up the building or other communal area. This, of course, rules out pre-payment coin-in-the-slot meters for security reasons.

D.C. mains supply

Direct-current is hardly ever used for distribution purposes, although there are still a few isolated d.c. systems. Usually, if a d.c. supply is required for particular equipment, e.g. for traction purposes, a rectifier is used to convert the a.c. supply to d.c. Special switchgear must be employed to control direct current, which is more susceptible to switching arcs.

Where d.c. systems still exist, they are usually transmitted on 3-wire cable networks. There are two 'main' conductors or 'outers' (positive and negative) and a neutral, or middle wire, which is usually earthed. The voltage between the positive main and the neutral is usually 220–250 V, as is the voltage between the negative and neutral. The voltage between positive and negative is normally double this. Normal domestic services are supplied across one main and neutral, while industrial loads are connected between positive and negative.

Control of circuits

Lighting circuits

Lighting circuits may be controlled from any number of positions. For example, hall, staircase or bedroom lighting is often conveniently controlled from two separate positions. This requires a two-way switch at each point and two extra wires, known as strapping wires, connecting these switches. The 'common' terminal on each switch can make contact with either strapping wire, but not both at once, so that one or other of the strapping wires carries the current when the circuit is closed (see Figs. 1.5, 1.6 and 1.7).

Extra control may be obtained by inserting one or more intermediate switches in the strapping wires. The intermediate switch allows the current to flow either along the uninterrupted length of the strapping wires, or connects these wires diagonally, thus shunting the current from one wire to the other.

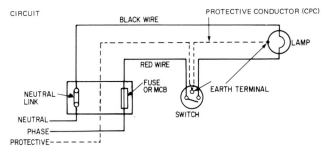

Fig. 1.5. Single light circuit fed from a consumer unit, and controlled by a single-pole switch. The protective conductor is required by IEE Regulations

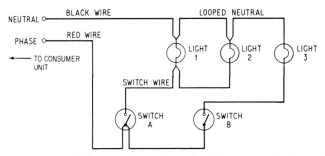

Fig. 1.6. Parallel lighting circuit. Lights 1 and 2 are controlled by switch A, while light 3 is controlled by switch B. Note that the protective conductor is not shown

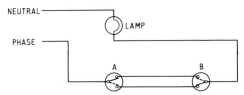

Fig. 1.7. Light controlled by two-way switches. This allows the light to be switched on or off from positions A and B. Note that the protective conductor is not shown

Further information on lighting circuits is given in Chapter **4**, while diagrams of basic circuits are given in Figs. 1.8, 1.9 and 1.10.

Ring Final circuits

A ring final circuit may be employed in accordance with the IEE Regulations for wiring socket-outlets employing fused plugs. The object is to provide

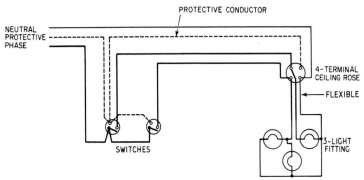

Fig. 1.8. A three-light fitting can be connected to a four-terminal ceiling rose using a flexible cord. It is controlled by two one-way switches to allow one, two or three lights to be switched on

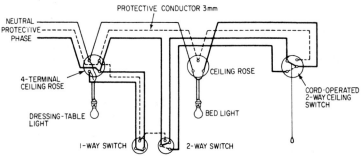

Fig. 1.9. Typical bedroom lighting circuit. The dressing table lisht is controlled by a one-way switch. The bed light is controlled by two two-way switches – one a wall mounted type and the other a cord-operated ceiling switch

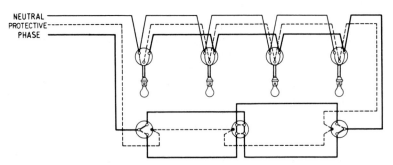

Fig. 1.10. Corridor or staircase lighting circuit, showing control from two-way switches and an intermediate switch

sockets in all positions where portable apparatus is likely to be required, thus avoiding the use of long flexible cables. Individual fixed appliances such as hand driers and inset fires may also be connected to a ring circuit, provided suitable fusing arrangements are used. It is not good practice to connect a fixed 3 kW water heater on a ring circuit. An example of a ring circuit is shown in Fig. 1.11 and its application is dealt with in Chapter 5.

Additional information on the installation of individual fixed appliances, inclusive of cooking, unrestricted and off-peak space-heating and water-heating appliances, is provided in Chapter 6.

Bell circuits

Bells may be operated by using a bell transformer connected to an a.c. supply or by using dry-cell batteries. The bell transformer usually has a secondary voltage of about 4 V, and thus the wiring to the bell circuits may have less insulation than that normally used for power and lighting circuits. Bell circuits are not normally run in the same conduit or in close proximity to power and lighting circuits but, if they are, the cable must be suitable for mains voltage.

Bell transformers should be wired either from a separate way in the consumer unit or from a fused connection unit in a lighting or power circuit. A 3 A fuse is normally adequate for domestic bell-circuit protection. Use suitable 240 V wiring from the fuse to the transformer, which may be mounted in any inconspicuous position. Extra-low voltage wiring may be used from the transformer to the push and bell positions. One wire is taken from the secondary terminals of the transformer to the bell push, and then from the push to one terminal of the bell. Another wire goes straight from the other terminal of the transformer to the bell.

An indicator board may be incorporated in bell circuits having a number of pushes, to show which has been operated, thus saving the cost of providing a separate bell for each push (see Fig. 1.12). The indicators may be of the flag type or lamp signals.

Intruder alarm systems

In recent years there has been a considerable increase in the installation of intruder alarm systems over a large range of premises. The types available are many, and encompass small domestic schemes to complex detection systems used in commercial property such as banks etc.

The days of electromechanical relays for use in control panels have long since passed, giving way to electronic and microelectronic systems. However, the principle of operation remains unchanged. An intruder activates a sensor which in turn operates an alarm sounder via a control panel.

There are two methods of wiring sensors or 'call points' (i) open circuit and (ii) closed circuit. Most control panels also cater for a mixture of both. Figs 1.13(a) and 1.13(b) illustrate the two methods.

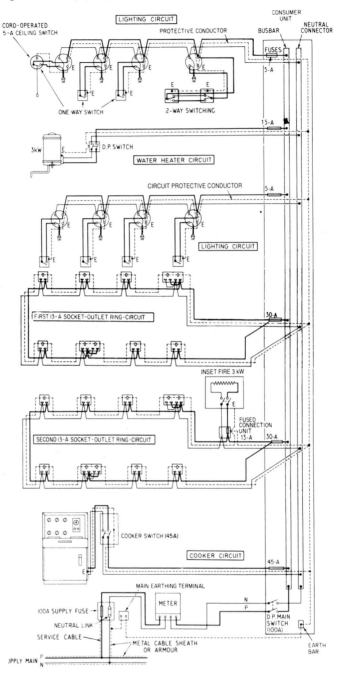

Fig. 1.11. A typical house distribution for lighting, heating and cooking. The single-phase service is from one phase and neutral of the mains supply

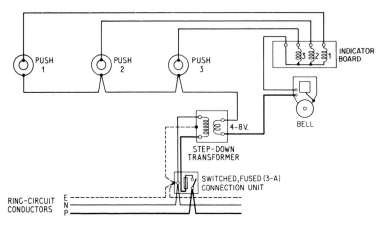

Fig. 1.12. Bell and indicator circuit. This may be supplied from any ring-circuit conductors using a fused connection unit and bell transformer

Typical call points include magnetic, door and window sensors, pressure mats, passive infra red and infra red beam, ultra-sonic sound sensors and vibration detectors etc., the choice of which being dependent on the particular environment.

(a)

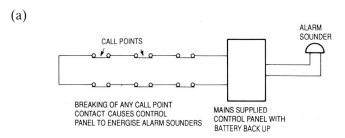

(b)

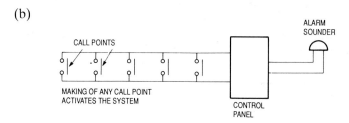

Fig. 1.13. Methods of wiring sensors or call points. (a) closed circuit. (b) open circuit

2 Ratings, cable sizes and circuit protection

All current-consuming electrical apparatus is given a rating which indicates the amount of power it will consume. This rating is in watts or kilowatts (kW), where 1 kW = 1000 watts. Some typical loadings of domestic apparatus are given below:

Appliance	Loading in watts
Lamps (filament)	15, 25, 40, 60, 75, 100, 150
Lamps (fluorescent)	15, 20, 30, 40, 50, 65, 75, 85
Fires and heaters	500–3400
Water-heaters, immersion heaters	750–3,000
Wash boilers	3,000
Washing machines	3,000
Spin-driers	300–500
Drying cabinets	2,000–3,000
Refrigerators	300–400
Cookers	6,000–8,000
Portable appliances (e.g. irons, vacuum cleaners, etc.)	10–3,000

Conductors and the current-carrying components of accessories (e.g. switches, fuses, MCBs, socket-outlets and plugs) must be large enough to carry the maximum current which the connected apparatus can cause to flow, without overheating or being overstressed. Conductors and accessories are rated in terms of current in amperes. therefore, before the required size or 'rating' of a conductor or accessory can be determined, the consumption of the connected apparatus in amperes must be calculated.

A resistance which allows 1 A of current to flow when connected across a 1 V power supply has a power consumption of 1 W. Thus:

$$\text{watts} = \text{amperes} \times \text{volts}$$

It follows from this expression that:

$$\text{amperes} = \frac{\text{watts}}{\text{volts}}$$

As an example, a 1 kW electric fire takes 4.16 A when connected to a 240 V

Table 2.1. Equivalent current and power ratings

Current	Voltage				
	100	200	230	240	250
(amperes)	(watts)	(watts)	(watts)	(watts)	(watts)
1	100	200	230	240	250
2	200	400	460	480	500
3	300	600	690	720	750
5	500	1,000	1,150	1,200	1,250
7	700	1,400	1,610	1,680	1,750
10	1,000	2,000	2,300	2,400	2,500
13	1,300	2,600	2,990	3,120	3,250
15	1,500	3,000	3,450	3,600	3,750

supply (1,000/240 = 4.16). Equivalent current and power ratings for different supply voltages are given in Table 2.1

The nominal output of an electric motor is normally expressed in kilowatts (kW) (0.75 kW is equivalent to 1 horsepower). It must be remembered however that this is an *output* value; the *input* to the motor will be higher due to the effects of efficiency and power factor. Details of the current taken by a motor should be obtained from the nameplate on the machine, or the manufacturers when necessary. Only very small motors are likely to be met in domestic installations.

Diversity factor

The size of a cable or accessory is not necessarily determined by the total power rating of all the current-consuming devices connected to it. It depends on what percentage of the connected load is likely to be operating at any one time, or the 'diversity factor'. IEE Regulation 311-01-01 states that 'In determining the maximum demand of an installation or parts thereof, diversity may be taken into account'.

Table 2.2 gives an indication of the maximum current that will normally flow in an installation, but it must be remembered that the figures given are only a guide. The amount by which the figures given are increased or decreased for any given installation should be decided by the engineer responsible for the design. The values given in the table refer to percentages of connected load or, where followed by the abbreviation 'f.l.', to the percentage of full-load current of a heating appliance, motor, socket-outlet or other current-using device. In calculating the maximum current, appliances and socket-outlets should be considered in order of their current ratings, the largest first.

Voltage drop in conductors

The relationship between the voltage and current in a circuit which has a certain resistance is given by Ohm's law as:

$$\text{current (amperes)} = \frac{\text{voltage (volts)}}{\text{resistance (ohms)}}$$

Table 2.2. Allowance for diversity

Estimation of the maximum current which will normally flow in an individual installation for use in computing the sizes of cables and equipment

Purpose of wiring	Estimation of maximum current
(1) Individual domestic installations, including flats of a block	
Lighting	66%
Heating and power appliances (but see exceptions below)	100% f.l. up to 10 A + 50% of any load in excess of 10 A
Cooking appliances permanently connected	10 A + 30% f.l. of connected cooking appliances in excess of 10 A; + 5 A if socket-outlet incorporated in unit
Water-heaters (instantaneous type)	100% f.l. of largest appliance; + 100% f.l. of second largest appliance; + 25% of remaining appliances
Water-heaters (thermostatically controlled)	No diversity allowable
Floor-warming installations	No diversity allowable
Thermal-storage space-heating installations	No diversity allowable
13 A socket-outlets and stationary appliances of rating not exceeding 13 A in radial and ring final circuits	100% current demand of largest circuit + 40% current demand of every other circuit
Socket-outlets and stationary appliances other than those listed above	100% f.l. of largest circuit + 40% f.l. of every other circuit

(2) Blocks of residential flats

Individual flats should be assessed in accordance with (1) above. The allowances for a block of flats should be assessed by a 'competent person'

(Based on IEE Regulations)

This means that, in all resistive circuits, a voltage drop is experienced, and for a given current the voltage drop increases with the length of the circuit.

IEE Regulations on voltage drop are as follows. **525-01-01**: Under normal service conditions the voltage at the terminals of any fixed current-using equipment shall be greater than the lower limit corresponding to the British Standard relevant to the equipment. Where the fixed current-using equipment concerned is not the subject of a British Standard the voltage at the terminals shall be such as not to impair the safe functioning of that equipment. **525-01-02**: The requirements of Regulation 525-01-01 are deemed to be satisfied for a supply given in accordance with the Electricity Supply Regulations 1988 (as amended) if the voltage drop between the origin of the installation (usually the supply terminals) and the fixed current-using equipment does not exceed 4% of the nominal voltage of the supply. A greater voltage drop may be accepted for a motor during starting periods and for other equipment with high inrush currents provided that it is verified that the voltage variations are within the limits specified in the relevant British Standards for the equipment or, in the absence of a British Standard, in accordance with the manufacturer's recommendations.

For practical purposes, voltage drop is often negligible in domestic wiring owing to the short distances or the ample size of the cables. However, with long runs of wiring, it is advisable to check that the drop in voltage from the

Table 2.3.
Current-carrying capacities (amperes):

Ambient temperature: 30°C
Conductor operating temperature: 70°C

Conductor cross-sectional area (mm²)	Reference Method 4 (enclosed in conduit in thermally insulating wall etc.)		Reference Method 3 (enclosed in conduit on a wall or in trunking etc.)		Reference Method 1 ('clipped direct')		Reference Method 11 (on a perforated cable tray horizontal or vertical)		Reference Method 12 (free air)		
	2 cables, single-phase a.c. or d.c. (A)	3 or 4 cables three-phase a.c. (A)	2 cables, single-phase a.c. or d.c. (A)	3 or 4 cables three-phase a.c. (A)	2 cables, single-phase a.c. or d.c. (A)	3 or 4 cables three-phase a.c. (A)	2 cables, single-phase a.c. or d.c. flat and touching (A)	3 cables, three-phase a.c. flat and touching or trefoil (A)	Horizontal flat spaced. 2 cables, single-phase a.c. or d.c. or 3 cables three-phase a.c. (A)	Vertical flat spaced. 2 cables, single-phase a.c. or d.c. or 3 cables three-phase a.c. (A)	3 cables, trefoil three-phase a.c. (A)
1	2	3	4	5	6	7	8	9	10	11	12
1	11	10.5	13.5	12	15.5	14	—	—	—	—	—
1.5	14.5	13.5	17.5	15.5	20	18	—	—	—	—	—
2.5	19.5	18	24	21	27	25	—	—	—	—	—
4	26	24	32	28	37	33	—	—	—	—	—
6	34	31	41	36	47	43	—	—	—	—	—
10	46	42	57	50	65	59	—	—	—	—	—
16	61	56	76	68	87	79	—	—	—	—	—
25	80	73	101	89	114	104	126	112	146	130	110
35	99	89	125	110	141	129	156	141	181	162	137
50	119	108	151	134	182	167	191	172	219	197	167
70	151	136	192	171	234	214	246	223	281	254	216
95	182	164	232	207	284	261	300	273	341	311	264
120	210	188	269	239	330	303	349	318	396	362	308
150	240	216	300	262	381	349	404	369	456	419	356
185	273	245	341	296	436	400	463	424	521	480	409
240	320	286	400	346	515	472	549	504	615	569	485
300	367	328	458	394	594	545	635	584	709	659	561
400	—	—	546	467	694	634	732	679	852	795	656
500	—	—	626	533	792	723	835	778	982	920	749
630	—	—	720	611	904	826	953	892	1138	1070	855
800	—	—	—	—	1030	943	1086	1020	1265	1188	971
1000	—	—	—	—	1154	1058	1216	1149	1420	1337	1079

Table 2.4.
Volt drops (mV/A/m)

		2 cables – single-phase a.c.										3 or 4 cables – three-phase a.c.										
Conductor cross-sectional area (mm²)	2 cables d.c. (mv)	Reference Methods 3 and 4 (Enclosed in conduit etc. in or on a wall) (mv)			Reference Methods 1 and 11 (Clipped direct or on trays, touching) (mv)			Reference Method 12 (Spaced*) (mv)			Reference Methods 3 and 4 (Enclosed in conduit etc. in or on a wall) (mv)			Reference Methods 1, 11, and 12 (In trefoil) (mv)			Reference Methods 1 and 11 (Flat touching) (mv)			Reference Method 12 (Flat spaced*) (mv)		
1	2	3			4			5			6			7			8			9		
		r	x	z	r	x	z	r	x	z	r	x	z	r	x	z	r	x	z	r	x	z
1	44	44			44			44			38			38			38			38		
1.5	29	29			29			29			25			25			25			25		
2.5	18	18			18			18			15			15			15			15		
4	11	11			11			11			9.5			9.5			9.5			9.5		
6	7.3	7.3			7.3			7.3			6.4			6.4			6.4			6.4		
10	4.4	4.4			4.4			4.4			3.8			3.8			3.8			3.8		
16	2.8	2.8			2.8			2.8			2.4			2.4			2.4			2.4		
25	1.75	1.80	0.33	1.80	1.75	0.20	1.75	1.75	0.29	1.80	1.50	0.29	1.55	1.50	0.175	1.50	1.50	0.25	1.55	1.50	0.32	1.55
35	1.25	1.30	0.31	1.30	1.25	0.195	1.25	1.25	0.28	1.30	1.10	0.27	1.10	1.10	0.170	1.10	1.10	0.24	1.10	1.10	0.32	1.15
50	0.93	0.95	0.30	1.00	0.93	0.190	0.95	0.93	0.28	0.97	0.81	0.26	0.85	0.80	0.165	0.82	0.80	0.24	0.84	0.80	0.32	0.86
70	0.63	0.65	0.29	0.72	0.63	0.185	0.66	0.63	0.27	0.69	0.56	0.25	0.61	0.55	0.160	0.57	0.55	0.24	0.60	0.55	0.31	0.63
95	0.46	0.49	0.28	0.56	0.47	0.180	0.50	0.47	0.27	0.54	0.42	0.24	0.48	0.41	0.155	0.43	0.41	0.23	0.47	0.40	0.31	0.51
120	0.36	0.39	0.27	0.47	0.37	0.175	0.41	0.37	0.26	0.45	0.33	0.23	0.41	0.32	0.150	0.36	0.32	0.23	0.40	0.32	0.30	0.44
150	0.29	0.31	0.27	0.41	0.30	0.175	0.34	0.29	0.26	0.39	0.27	0.23	0.36	0.26	0.150	0.30	0.26	0.23	0.34	0.26	0.30	0.40
185	0.23	0.25	0.27	0.37	0.24	0.170	0.29	0.24	0.26	0.35	0.22	0.23	0.32	0.21	0.145	0.26	0.21	0.22	0.31	0.21	0.30	0.36
240	0.180	0.195	0.26	0.33	0.185	0.165	0.25	0.185	0.25	0.31	0.17	0.23	0.29	0.160	0.145	0.22	0.160	0.22	0.27	0.160	0.29	0.34
300	0.145	0.160	0.26	0.31	0.150	0.165	0.22	0.150	0.25	0.29	0.14	0.23	0.27	0.130	0.140	0.190	0.130	0.22	0.25	0.130	0.29	0.32
400	0.105	0.130	0.26	0.29	0.120	0.160	0.20	0.115	0.25	0.27	0.12	0.22	0.25	0.105	0.140	0.175	0.105	0.21	0.24	0.100	0.29	0.31
500	0.086	0.110	0.26	0.28	0.098	0.155	0.185	0.093	0.24	0.26	0.10	0.22	0.25	0.086	0.135	0.160	0.086	0.21	0.23	0.081	0.28	0.30
630	0.068	0.094	0.25	0.27	0.081	0.155	0.175	0.076	0.24	0.25	0.08	0.22	0.24	0.072	0.135	0.150	0.072	0.21	0.22	0.066	0.28	0.29
800	0.053	–			0.068	0.150	0.165	0.061	0.24	0.25	–			0.060	0.130	0.145	0.060	0.21	0.22	0.053	0.28	0.29
1000	0.042	–			0.059	0.150	0.160	0.050	0.24	0.24	–			0.052	0.130	0.140	0.052	0.20	0.21	0.044	0.28	0.28

Spacings larger than those specfied in Method 12 will result in larger volt drops.

supply terminals to the most distance appliance will be within the prescribed limit, and if necessary to use a larger size of cable. It is important that the IEE recommendations be adhered to, since all electrical apparatus suffers a loss of power and efficiency when supplied with less than the required voltage.

In order to check the voltage drop in each section of the wiring, reference should be made to Tables 2.4 and 2.6. These give the voltage drop per ampere per metre for cables likely to be used in domestic wiring.

Cable sizes

The ratings of cables and flexible cords are given in the IEE Tables 4D1A to 4L4A. Tables 2.3, 2.5 and 2.7 are based on these.

In using the tables, current ratings are to be modified where applicable by rating factors in respect of: (1) Ambient temperature, (2) Class of excess-current protection, (3) Grouping, (4) Disposition (5) Thermal insulation. When the installation conditions differ in more than one respect from those given in the appropriate table, a separate factor for each special circumstance must be applied. The appropriate factors are to be applied as multipliers to the current ratings or alternatively divided into the rating of the protective device. Details are given in IEE Appendix 4.

1 Ambient temperature
The current ratings given in the tables for cables and bare conductors are based on an ambient air temperature of 30°C. Where the ambient air temperature exceeds this figure, the appropriate rating factor must be applied, giving a lower rating. On the other hand, if it can be established that the ambient temperature will not exceed 25°C, use may be made of the rating factor appropriate to this temperature, which increases the current rating.

2 Class of excess-current protection
Certain thermoplastic materials (e.g. PVC) deteriorate if subjected to sustained high temperatures. Therefore, the current rating of cables insulated with PVC or synthetic rubbers are determined not only by the maximum conductor temperature admissible for continuous running but also by the probable duration of excess current.

The current ratings in Appendix 4 of the IEE Regulations are based on the cable being protected by a cartridge fuse to BS 88 or BS 1361, or circuit-breaker to BS 3871 or BS 4752. If, however, the cable is protected by a semi-enclosed (rewirable) fuse to BS 3036, the current ratings stated must, for the tables indicated, be reduced by applying the factor 0.725. This is due to the longer operating time of semi-enclosed fuses.

3 Grouping
The ratings given in the tables are for single circuits only. For groups, the appropriate rating factor must be applied.

4 Disposition
The ratings given in the tables also apply only to the dispositions stated. Rating factors for cables installed in enclosed trenches, are given in the IEE Regulations, Appendix 4.

Table 2.5.
Volt drops (mV/A/m)

Conductor cross-sectional area (mm²)	Reference Method 4 (enclosed in an insulated wall, etc.)		Reference Method 3 (enclosed in conduit on a wall or ceiling, or in trunking)		Reference Method 1 (clipped direct)		Reference Method 11 (on a perforated cable tray), or Reference Method 13 (free air)	
	1 two-core cable* single-phase a.c. or d.c. (A)	1 three-core cable* or 1 four-core cable, three phase a.c. (A)	1 two-core cable* single phase a.c. or d.c. (A)	1 three-core cable* or 1 four-core cable, three phase a.c. (A)	1 two-core cable* single-phase a.c. or d.c. (A)	1 three-core cable* or 1 four-core cable, three-phase a.c. (A)	1 two-core cable* single phase a.c. or d.c. (A)	1 three-core cable* or 1 four-core cable, three-phase a.c. (A)
1	2	3	4	5	6	7	8	9
1	11	10	13	11.5	15	13.5	17	14.5
1.5	14	13	16.5	15	19.5	17.5	22	18.5
2.5	18.5	17.5	23	20	27	24	30	25
4	25	23	30	27	36	32	40	34
6	32	29	38	34	46	41	51	43
10	43	39	52	46	63	57	70	60
16	57	52	69	62	85	76	94	80
25	75	68	90	80	112	96	119	101
35	92	83	111	99	138	119	148	126
50	110	99	133	118	168	144	180	153
70	139	125	168	149	213	184	232	196
95	167	150	201	179	258	223	282	238
120	192	172	232	206	299	259	328	276
150	219	196	258	225	344	299	379	319
185	248	223	294	255	392	341	434	364
240	291	261	344	297	461	403	514	430
300	334	298	394	339	530	464	593	497
400	—	—	470	402	634	557	715	597

Table 2.6.
Volt drops (mV/A/m)

Conductor cross-sectional area	Two-core cable dc	Two-core cable single phase a.c.			Three- or four-core cable three phase a.c.		
1	2	3			4		
mm²	mV	mV			mV		
1	44	44			38		
1.5	29	29			25		
2.5	18	18			15		
4	11	11			9.5		
6	7.3	7.3			6.4		
10	4.4	4.4			3.8		
16	2.8	2.8			2.4		
		r	x	z	r	x	z
25	1.75	1.75	0.170	1.75	1.50	0.145	1.50
35	1.25	1.25	0.165	1.25	1.10	0.145	1.10
50	0.93	0.93	0.165	0.94	0.80	0.140	0.81
70	0.63	0.63	0.160	0.65	0.55	0.140	0.57
95	0.46	0.47	0.155	0.50	0.41	0.135	0.43
120	0.36	0.38	0.155	0.41	0.33	0.135	0.35
150	0.29	0.30	0.155	0.34	0.26	0.130	0.29
185	0.23	0.25	0.150	0.29	0.21	0.130	0.25
240	0.180	0.190	0.150	0.24	0.165	0.130	0.21
300	0.145	0.155	0.145	0.21	0.135	0.130	0.185
400	0.105	0.115	0.145	0.185	0.100	0.125	0.160

Table 2.7

Nominal cross-sectional area of conductor (mm²)	Maximum diameter of wires forming conductor (mm)	Current-carrying capacity Single-phase a.c. (A)	Three-phase a.c. (A)	Maximum mass supportable by twin flexible cord (ng)
1	2	3	4	5
0.5	0.21	3	3	2
0.75	0.21	6	6	3
1	0.21	10	10	5
1.25	0.26	13	–	5
1.5	0.26	16	16	5
2.5	0.26	25	20	5
4.0	0.31	32	25	5

Correction factor for ambient temperature:

60°C rubber and PVC cords

Ambient temperature	35°C	40°C	45°C	50°C	55°C
Correction factor	0.92	0.82	0.71	0.58	0.41

85°C rubber cords having a hofr sheath or a heat-resisting PVC sheath

Ambient temperature	35°C to 55°C	60°C	65°C	70°C
Correction factor	1.0	0.96	0.83	0.67	0.47

150°C rubber cords

Ambient temperature	35°C to 120°C	125°C	130°C	135°C	140°	145°C
Correction factor	1.0	0.96	0.85	0.74	0.60	0.42

Glass fibre cords

Ambient temperature	35°C to 150°C	155°C	160°C	165°C	170°C	175°C
Correction factor	1.0	0.92	0.82	0.71	0.57	0.40

5 Thermal insulation

If a cable is surrounded by thermal insulation for 0.5 m or more, its rating must be reduced by factor of 0.5.

Example of cable sizing

To determine the minimum cable size (PVC twin) to supply a load of 28 A with a cable length of 40 m, protection by BS88 fuses and no correction factors, we must follow a set procedure:

1   Design current of circuit $I_B = 28$ A
2   Select rating of protection I_N, such that $I_N \geqslant I_B$. Hence $I_N = 30$ A
3   Select correction factors (CF) (not applicable)
4   Determine current carrying capacity of cable I_Z from $\dfrac{I_N}{CF}$

hence $I_Z = I_N = 30$ A

5   Choose cable to suit I_Z from tables, i.e. 4.0 mm²
6   Check volt drop from:

$$VD = \frac{mV \times I_B \times length}{1000}$$

$$= \frac{11 \times 28 \times 40}{1000} = 12.32 \text{ V (too high)}$$

Hence try 6.0 mm² cable

$$VD = \frac{7.3 \times 28 \times 40}{1000} = 8.2 \text{ V}$$

This is acceptable, as 4% of 240 V is 9.6 V.
Hence, due to volt drop constraints, a 6.0 mm² cable would be needed.

Fuses

All electrical circuits must be protected against excessive current. If current in excess of the rating of the circuit were allowed to go on indefinitely, damage would obviously result. Fuses (or over-current circuit-breakers) are installed in such a manner that the faulty or overloaded section of the circuit is disconnected from the supply before any damage occurs. The over-current devices are so constructed that a large overload causes the circuit to be opened in a very short time, while for a small overload a correspondingly longer time elapses before the fuse or circuit-breaker operates.

A fuse comprises all the parts which form the complete device — holder, link, contacts and contact base. The fuse element is the part which melts and opens the circuit, and the fuse link is the part which contains the element.

The current rating of a fuse is the maximum current that the fuse will carry without exceeding a specified temperature rise. A fuse of a given current rating is designed to carry that current continuously without overheating and to protect a circuit of the same nominal rating. The fuse element actually melts or 'blows' at a higher current than the current rating of the fuse.

The minimum fusing current at which the fuse element will melt divided by the current rating is known as the 'fusing factor' of the fuse. For example, a 15 A semi-enclosed fuse will blow at 29 A, giving a fusing factor of 29/15, i.e. just below 2.0. As mentioned earlier the type of over-current protection affects the value of current rating applied to the cables concerned.

Breaking capacity

It is necessary from the point of view of safety that when the fuse link breaks, arcing energy does not cause mechanical damage to the fuse carrier, fuse contacts or case. The nearer the circuit to the source of supply the greater the current that will pass through it when a short circuit occurs. The fuse must therefore be capable of safely interrupting the greatest possible short-circuit current that may pass.

The ordinary semi-enclosed fuse has a limited breaking capacity owing to the difficulty of extinguishing the arc which takes place with high fault current. A great improvement in breaking capacity is obtained by using cartridge fuses and, for very high fault currents, cartridge fuses of high-breaking-capacity (h.b.c.) are employed.

Semi-enclosed (rewirable) fuses

The simplest type of fuse in general use is the semi-enclosed (rewirable) type, consisting of a base and carrier of an incombustible material, such as porcelain or moulded plastic. The base encloses the fixed contacts to which the incoming and outgoing cables are connected. The link or carrier has two contacts, connected by means of screws or nuts to the fuse wire. The fuse wire is normally enclosed in the porcelain link (see Fig. 2.1), or in an asbestos tube, to prevent damage being caused by the arc and the resulting hot gases when

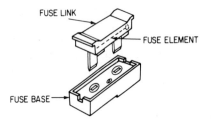

Fig. 2.1. Semi-enclosed fuse

the fuse wire melts. A disadvantage of semi-enclosed fuses, other than their limited breaking capacity and longer operating time, is the fact that a blown fuse wire may be replaced by one of an incorrect size.

Cartridge fuses

The fuse consists of a tube with two end caps to which the fuse element is attached, and a filler. The tube or body is normally made of porcelain, or bakelite, of high strength and resistance to intense local heating. The end caps, of tinned brass or copper, are attached rigidly to the body of the fuse so that they do not blow off when the fuse operates. With the larger sizes of fuse, lugs or tags are brazed to the end caps for attaching the cartridge to the fuse carrier. The elements are either wire or strip metal, often with a 'neck', or a section of low melting point in the centre. The cartridges are filled with quartz sand or other suitable material to absorb and extinguish the arc. The advantages of cartridge fuses are:

(a) the possibility of obtaining high rupturing capacity;
(b) the fuses, including those with high current ratings, are very compact, do not overheat or deteriorate in service and operate noiselessly and without arcing;
(c) the declared rating is accurate and consistent;
(d) operation is rapid, and operating time is inversely proportional to fault current; and
(e) the small dimensions and no external arcing enable cartridge fuses to be used in plugs, small fuse carriers and distribution boards.

Fuse sizes

The correct sizes of tinned copper and standard alloy fuse wires are given in Table 2.8. Current ratings of cartridge fuses are marked on the fuse and also indicated by the colour of the fuse body.

Renewing a fuse

The fuse wire or cartridge which is used to replace the one which has 'blown' should always have the same current rating as the one replaced. Before replacing a fuse, it is advisable to first check the circuit involved to ascertain the reason for the fuse blowing and isolate the faulty section if possible. The switch controlling a fuseboard should always be switched off before removing or replacing fuse carriers.

Renewing a cartridge fuse is simply a matter of removing the old one from the holder and replacing it with a new one of the same rating.

To renew a semi-enclosed fuse it is necessary to remove any sections of old fuse wire and rethread the carrier with fuse wire of the same rating. Care must be taken not to overtighten the screws, or stretch the fuse wire too tightly between them, as both reduce the cross-sectional area of the wire and consequently the current-carrying capacity. Fuse carriers are normally marked with the maximum rating of fuse wire to be fitted.

Table 2.8. Sizes of fuse elements of plain and tinned copper wire for use in semi-enclosed fuses

Nominal fuse current (A)	Nominal diameter of wire (mm)
3	0.15
5	0.20
10	0.35
15	0.50
20	0.60
25	0.75
30	0.85
45	1.25
60	1.53
80	1.80
100	2.00

Note: The use of cartridge fuses to the appropriate British Standard is normally recommended, but where a semi-enclosed fuse is used, the figures given in this table will, in the absence of recommendations made by the maker of the fuse, provide an approximate guide to the size of wire required. These figures represent the current which the fuse will carry continuously without overheating; the value at which the fuse will blow is approximately twice this, depending upon the type and construction of the fuse.

Miniature circuit-breakers

As an alternative to fuses, small automatic circuit-breakers are available which can be mounted in a distribution board or consumer unit in much the same way as fuses. These miniature circuit-breakers (MCBs) are, in fact, switches designed to open automatically when the current passing through them exceeds the value for which they are set. MCBs normally have a time-delay-type tripping characteristic, the operating time being controlled by the magnitude of the overcurrent. Thus, the circuit-breaker is not affected by transient overloads, such as motor-starting currents and switching surges, and will operate only if the overload is present long enough to constitute a danger to the circuit being protected.

Miniature circuit-breakers can perform many functions. They can be used as local control switches, or to prevent installations, appliances or equipment from being overloaded. They may also be used as fault-making isolating switches — that is, isolating switches capable of making and breaking rated current and also of withstanding reclosure against an existing short-circuit fault.

MCBs are normally installed to give over-current protection, and IEE Regulations (Chapter 43) state that they must operate at a current not exceeding 1.45 times the designed load current of the circuit. MCBs that meet the requirements of IEE Appendix 4 and Regulation 433-02-01 may be deemed to comply with the foregoing. MCBs have certain advantages over

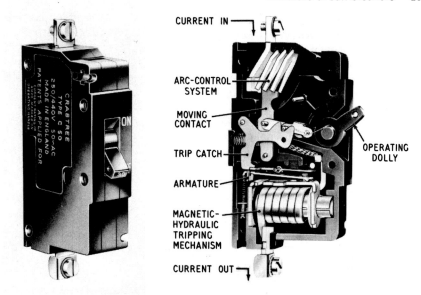

CURRENT IN

ARC-CONTROL
SYSTEM

MOVING
CONTACT

TRIP CATCH

ARMATURE

MAGNETIC-
HYDRAULIC
TRIPPING
MECHANISM

CURRENT OUT

OPERATING
DOLLY

Fig. 2.2. Miniature circuit-breaker (*Crabtree Electrical Industries*)

fuses, and the over-current protection provided may allow economies in cable sizes. Also, they do not 'age' in service and since their settings cannot be altered, the risk of over-loading of circuits due to unqualified persons operating them is completely removed.

When a circuit-breaker operates, or 'trips', a visual indication is immediately given since the switch dolly moves automatically to the off position. Once the fault has been traced and rectified or isolated, the circuit-breaker can simply be switched on again. Even if the circuit-breaker is closed with the fault still existing and the switch dolly is held or wedged in position, it will still switch itself off automatically (see Fig. 2.2).

Automatic operation and the time-delay characteristic can be obtained in two ways, magnetic tripping or thermal tripping.

Magnetic tripping

The current through the circuit-breaker passes through an electromagnet which has a pivoted metal arm at a set distance from one end. At normal running current, the magnetic pull is not sufficient to attract the arm, but when overloading occurs the magnetic pull is increased, the arm is attracted to the magnet and trips the switch. Time-delay effect can be achieved by filling the centre of the electromagnet coil with a silicone damping fluid and inserting a movable metal slug held in position by a light spring.

Under normal running conditions, the slug is held in position by the light spring, but with overload currents the magnetic pull exceeds the restraining force of the spring (see Fig. 2.3). The slug therefore moves towards the other end of the coil, its speed of travel being controlled by the damping fluid and the magnitude of the current. Only when the slug reaches the far end of the coil is the magnetic field strong enough to attract the moving arm and trip the circuit-breaker. Heavy overloads or short circuits produce a sufficiently strong magnetic field to attract the arm irrespective of the position of the slug. This type of tripping mechanism can be immediately reset after tripping.

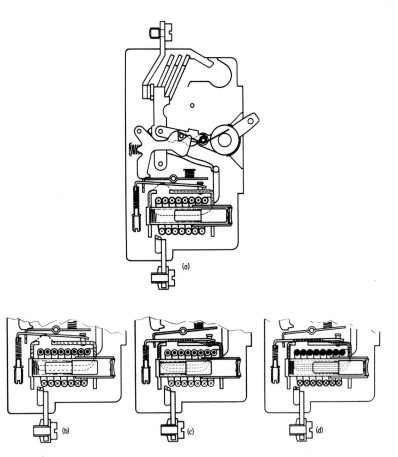

Fig. 2.3. Magnetic tripping of a circuit-breaker. Under normal conditions the iron slug is at far end of the tube (a). When an overload occurs the magnetic pull from the trip coil causes the slug to move (b), until it trips the circuit breaker (c). For a heavy overload or short circuit, the greater magnetic pull on the armature is sufficient to trip the circuit-breaker instantaneously without the slug moving from the rest position (d).

The magnetic-hydraulic method of time-delay tripping illustrated is that employed in the Crabtree Type C-50 MCB

Thermal tripping

Automatic operation with a time-delay characteristic is provided in this case by means of a thermally operated bi-metallic element. When the bi-metallic element is heated to a given temperature, the resulting deflection is arranged to trip the circuit-breaker, the time-delay characteristic being provided by the time taken to heat the element to this temperature. The element may carry the line current and thus be directly heated, or a separate heater may be used. With this type of circuit-breaker, instantaneous operation under short-circuit or heavy-overload conditions is provided by using an additional tripping element. This is normally of the electromagnetic type and either trips the switch itself or helps to speed the bending of the bi-metal (see Fig 2.4).

Thermal-type units are affected by ambient temperature and also need to

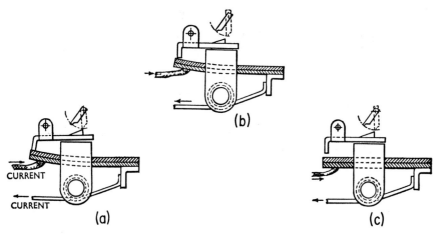

Fig. 2.4. Combined thermal and magnetic tripping mechanism showing conditions for (a) small overloads, when only the bi-metal trips the mechanism; (b) heavier overloads, when both bi-metal and magnetic coil trip the mechanism; and (c) heavy overloads and short-circuits, when magnetic coil acts alone

be left for a short time after tripping to enable the bi-metal and heater to cool down before being reclosed.

Triple-pole circuit-breakers

Triple-pole circuit-breakers can be used to protect 3-phase equipment and installations, thus ensuring that a fault on any one phase causes all three phases to be tripped out. This avoids the possibility of damage being caused to equipment or motors through 'single-phasing'.

3 Domestic installation practice

When an electricity supply to a new installation, or an appreciable extension to an existing installation is proposed, the local electricity board should first be consulted. They should be given details of the prospective load, as reinforcement or extension to their mains may be necessary and they may ask for a contribution towards the cost of this. The board will indicate whether a new supply will be overhead or underground. The point of entry into the building will also require to be discussed and agreed. They will also advise on the most suitable type of metering and the tariff to be adopted.

For a new supply, the electricity board will provide either a lead- or plastic-covered underground cable or overhead wires to the building, with PVC/PVC-insulated leads to their main fuse position. For normal domestic installations, a single-phase supply is usually provided. This means one phase wire, a neutral wire and, with underground services and certain types of overhead services, a separate earth wire. The phase wire is taken directly to the board's main cutout, which contains their main fuse. The neutral and earth wires are terminated on separate connecting blocks.

From the cutout and neutral block two insulated leads are taken by the board to the meter. 'Tails' are provided by the consumer from his main switch or consumer unit to the meter. These must have a current rating not less than that of the board's main fuse and if unsheathed cables are used they must be installed in suitable trunking or conduit. Details should be agreed with the board.

For exceptionally heavy loads the incoming supply may be 2- or 3-phase, and an additional fuse is then installed by the board in each phase wire.

Main switches

IEE Regulations Chapters 13 and 476 require the provision of either a linked switch or a linked circuit-breaker, which is arranged to disconnect all live circuit conductors of each installation from the supply.

Each installation that is to be metered on a different tariff must have a separate main switch and distribution fuseboard or consumer unit to isolate the whole of that installation, e.g. off-peak heating and installations with separate tariffs for power and lighting. A separate means of isolation should be provided in each detached building that may be connected to a consumer's installation.

Excess current protection

IEE Regulations require a means of over-current protection, comprising a fuse or circuit-breaker, in each phase conductor of the supply. This may be omitted, however, provided that the rating of all cables connected between the electricity board's fuse or circuit-breaker and the consumer's fuses or circuit-breakers is not less than the rating of the board's fuse or circuit-breaker.

Consumer units

For domestic installations it is usual to install a consumer unit (or units) adjacent to the meter position.

A consumer unit consists of a main switch, usually rated at 60, 80 or 100 A and the required number of outgoing circuits, or 'ways'. These ways may be rated at 5, 15, 30 or 45 A or 6 A, 16 A, 32 A etc. in various combinations, and may consist of semi-enclosed or cartridge fuses, or MCBs. Units with up to 14 outgoing ways are available.

On some units the main switch is replaced by a residual-current earth-leakage circuit-breaker (see Chapter 10) which provides earth-leakage protection on outgoing circuits and also acts as a main isolator for the consumer unit. Fig. 3.1(a) shows a typical standard consumer unit equipped with MCBs.

'Split-load' consumer units are also available on which some of the outgoing ways are fed via a common ELCB and the other ways are fed direct from the main switch; however the main switch controls *all* the outgoing circuits. Fig. 3.1(b) shows a typical unit of this type having four ways fed via the ELCB and three ways direct from the main switch. These units permit some circuits, e.g. socket-outlet ring circuits, to have earth-leakage protection whilst others, such as lighting, are fed direct.

When two tariffs are being used, e.g. normal and off-peak, it is necessary to install two consumer units, one for each tariff. These can now be combined in a single 'twin-tariff' mounting which is, in effect, two separate consumer units fitted in a common case (Fig. 3.1(c) and (d)). Each bank of outgoing circuits is controlled by a separate main switch fed by a separate mains supply.

Consumer units suitable for 2- or 3-phase supplies are also available but these are not very common on domestic work.

Various combinations of the above options are possible and in special cases the manufacturer should be consulted.

Consumer units include neutral and earth bars and a main earthing terminal. Units are available for flush or surface mounting and the case may be metalclad or insulated, the former being used for conduit systems. With a sheathed wiring system an insulated consumer unit is usual, but a metalclad unit may be used if desired.

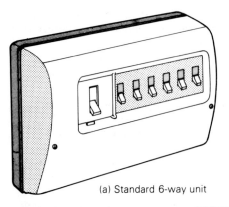

(a) Standard 6-way unit

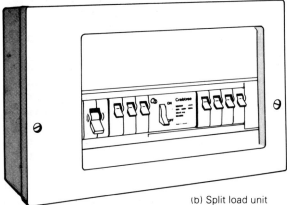

(b) Split load unit

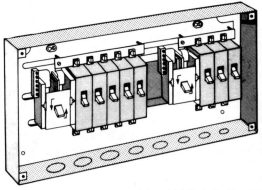

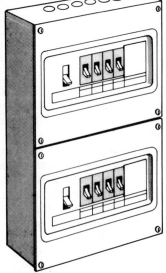

(c) and (d) Twin tariff units

Fig. 3.1. Typical consumer units
(*Crabtree Electrical Industries*)

The wiring for a domestic cooker should be connected on a separate circuit, the rating being determined in accordance with the rules for diversity given in Chapter 2, Table 2.2. Large cookers may require a 45 A way.

Ring circuits for socket-outlets to BS 1363 have both ends of the phase and neutral wires brought back to the same fuse (or MCB) and neutral terminal respectively in the consumer unit. Each separate ring circuit must be protected by its own 30 A or 32 A fuse or MCB. Radial circuits for socket-outlets to BS 1363 must be fed via 20 A or 30 A or 32 A fuses or MCBs, the rating depending on the cable size used. Further details are given in Chapter 5.

Two lighting circuits are normally provided, connected to two separate 5 A or 6 A ways. This ensures that the whole dwelling will not be in darkness should a fault occur on one lighting circuit.

When the wiring is brought into the consumer unit the neutral wires should be connected in the same order as their respective phase wires.

It is customary for the electricity board to install cutouts and meters on a wall-mounted wood or fibre board, porcelain spacers or wood battens being used to give approximately 25 mm clearance between the back of the board and the wall. This board is the property of the electricity board and normally no other equipment may be attached to it. The consumer unit should be mounted on a separate board adjacent to the meter board. This prevents rusting on the back of metal enclosures and permits sheathed wiring concealed under the plaster to be brought in the back of the unit. Where conduit systems are employed, top entry is usually more convenient.

Conduit and cable entries

All consumer units are provided with knockouts — stamped sections of the case which can be tapped out with a hammer to admit cables or conduit. Conduit is normally terminated in metalclad fuseboards with a smooth brass bush, to prevent abrasion of cables. Where sheathed cables enter metalclad units, a rubber grommet or brass bush with locknut must be inserted for this purpose. Fuseboards and consumer units must be totally enclosed, to prevent

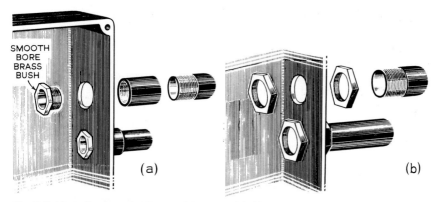

Fig. 3.2. Methods of terminating conduit at metalclad boxes
  (a) Smooth bore bush and coupling
  (b) Locknut

the ingress of dirt or moisture or accidental touching of live parts; therefore, grommets should be tight fitting and any unused conduit entries should be blanked off.

There must be a good mechanical and electrical connection between metallic conduit and metalclad fuseboards. This is usually achieved by screwing the end of the conduit and fitting a brass bush and locknut as shown in Fig. 3.2. If there is a break between metal casings of cables and main switch or distribution chambers, bonding clips or clamps and copper wires must be used to ensure proper bonding and earthing.

Consumer units should be clearly labelled so that the circuits may be identified easily.

Switchfuses and consumer units

It is sometimes an advantage and cheaper for very small domestic installations, or extensions to existing installations, to be controlled by means of switchfuses or splitter units. A circuit requiring separate metering, e.g. off-peak heating, may also be controlled by this method (see Chapter 6). These items are really small consumer units. Switchfuses comprise a single- or double-pole main switch and one fuse or MCB, while consumer units have two or more fuses or MCBs. The casing may be metal or insulated and is normally completely detachable from the back plate. It is not normally possible to remove the casing until the switch is 'off'.

As with consumer units, switchfuses are provided with knockouts for conduit or cable entry, and both metalclad and insulated types are normally provided with a screw or clamp for earth connection. They should also be installed on a wooden board, in a similar manner to consumer units.

With all-insulated units, provision must be made to bond together and earth any protective conductors or metal covering of cables.

Connection of cables

After removing the insulation from the ends of the cables, the end of each conductor, if stranded, should be twisted together for connection to the appropriate terminal. Cable sockets should be sweated to the conductors if their section is 25 mm² or more (or less if the equipment terminals will allow it). A little slack cable should be left after connection to the terminal is made, to allow renewal of switches etc.

Branch switches

With normal domestic supplies where the neutral conductor is connected to earth, single-pole switches (i.e. switches which make-and-break only one conductor of the supply) may be used to control lighting points or fixed equipment, although there is no objection to the use of double-pole switches. It is recommended that double-pole switches should be used for fixed heating appliances where the heating elements can be touched (Chapter 6).

On domestic installations, switches that control individual or collective lighting points are usually 5 A single-pole. Socket-outlets and fused connection units that incorporate a switch are dealt with in Chapter 5.

Switches for wall mounting may be of the tumbler or rocker type, the latter being more common, and may be surface, semi-recessed or flush pattern. They are available in single- or multi-gang units.

Ceiling mounted switches, operated by shockproof cords are an alternative to wall switches and these are available in surface or semi-recessed patterns. They are often used in bathrooms when a wall-mounted switch is not permissible or in other locations when there is difficulty in wiring to a wall-mounted switch.

Tumbler-operated switches have a protruding dolly with a positive up-and-down movement (see Fig. 3.3), whereas rocker-type switches have a centre-pivoted dolly which closes the micro-gap switch contacts when the bottom is depressed and opens them when the top is depressed (see Fig. 3.4). This type

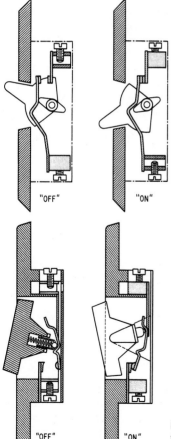

"OFF"   "ON"

Fig. 3.3. Tumbler operated lighting switch showing cam-operated moving contact and operation of switch dolly

"OFF"   "ON"

Fig. 3.4. Rocker-operated switch mechanism, showing moving contact and spring-loaded ball and socket operating mechanism

of switch is extremely easy to operate, and less likely to be damaged if accidentally knocked.

A suitable mounting height for wall switches controlling lighting points is 1.37 m above floor level. A uniform height is desirable.

Switches in dangerous situations

IEE Regulations Part 6 requires that any switches in a room containing a fixed bath or shower must be situated so as to be normally in accessible to a person using a fixed bath or shower. In such cases, the switch may be placed in a suitable position outside and immediately adjacent to the normal access door of the room, or may be of the type (described above) operated by an insulating cord (Fig. 3.5).

Fig. 3.5. Cord-operated 5 A ceiling switch

In damp situations, all switches must be of the damp- and dust-proof type, and cable entries must be provided with glands or bushings or be suitable to receive screwed conduit.

Special requirements apply in situations where there is likely to be flammable or explosive dust, vapour, gas or liquids. Guidance on this is given in Code of Practice CP 1003 but it is advisable to obtain expert advice. However such cases are not very common on domestic work.

Plateswitches and gridswitches

Switches are manufactured in two basic designs, either plateswitch or gridswitch type.

Plateswitches are constructed with the switch mechanism, complete with

Fig. 3.6. Two-gang gridswitch. The slotted holes are for alignment (*MK Electric Limited*)

switch dolly, permanently attached to the back of the surface plate. They can be either rocker- or tumbler-type and may be constructed of moulded-plastic materials throughout or have a metal surface plate. They are available in multi-gang units up to about six-gang. The one-gang type can be either one-way or two-way but multi-gang units normally have two-way switches. Two-way switches can, of course, be connected as one-way if required. The terminals on 5 A switches will normally accept two 1.5 mm² cables.

The gridswitch system uses interchangeable switches and components which are first assembled on grids. The switch mechanism either clips or is screwed on to the grid, which in turn is screwed into a recessed box. The cover plate is then screwed on to the grid. There is normally a levelling and squaring arrangement on the grid to allow for different plaster depth and box settings. Boxes for use with gridswitches are often suitable for either surface or flush mounting, and a choice of metal or insulated switchplates up to 24-gang, with suitable boxes, is available.

Most manufacturers produce a full range of interchangeable units, such as rocker or tumbler switches, bell-push switches, neon indicators and dimmer switches, which all fit onto the grid in the same manner, thus making alterations and additions extremely easy. The units can be wired before insertion into the grid, and hence connection is very quick and simple (see Fig. 3.6).

Fitting flush switches

Flush mounting of wall switches is the most popular method in modern domestic wiring. Any of the usual systems used for domestic wiring may be adapted for the connection of flush switches.

The flush switch must be fixed in a box let into the wall so that, on completion of the plastering, the box edges are level with or below the plaster surface. The wiring to the box is buried in the plaster, and should preferably be enclosed in metal or plastic conduit or channelling.

Boxes for flush switches may be of cast-iron, pressed-steel, moulded plastic or hardwood and are available for housing single, double or multi-gang switches. Metal boxes are essential where the conduit is used as the protective conductor for through type boxes or earthed switches.

An earthing terminal, connected to the protective conductor of the final circuit, must be provided at every lighting-switch position, unless this takes the form of an earthed metal box having a means of fixing the switchplate in reliable electrical contact with the box. The protective conductor must be taken to all lighting switch positions.

Metal boxes may be obtained, already drilled and tapped for switch-fixing screws, to suit the dimensions of the switches to be employed. The boxes are also obtainable with various types of entry: rubber-bushed, tapped, spout, knockout or lug-grip to suit the wiring system adopted. The usual size of terminal-box entry for conduit is 20 mm.

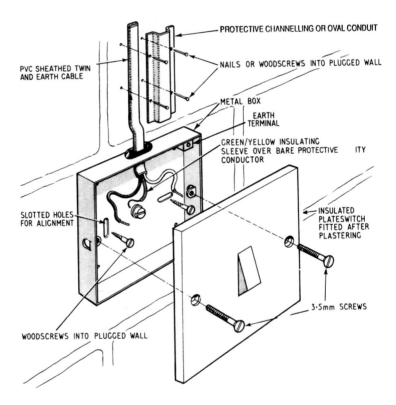

Fig. 3.7. Cable entering a flush switch can also be protected by steel or PVC channelling with a rubber grommet at entry position. An earth terminal must be provided and an insulating sleeve placed over the bare protective conductor

To accommodate the box so that it does not project beyond the plaster level, the wall may need to be cut out to a depth to suit the depth of the box and the thickness of the plaster layer. Boxes are obtainable in depths of 16, 25, 35 and 46 mm. The box is selected according to the system of wiring and the depth of the plaster. A deep box must be used for large-diameter conduit and a shallower box may be used for concealed PVC cable. The box is attached to the wall, after the brickwork has been chiselled out to the appropriate depth, when required, by drilling and plugging the wall and using short wood screws. Ideally the depth of the box should be slightly less than the depth of the plaster; the box can then be fixed direct to the brickwork without chiselling it out.

The wiring is pulled into the box and enough slack cable left to connect the switch (see Fig. 3.7). Where sheathed-wiring systems are employed, an earth terminal must be provided at the switch position, and this will normally be on the side of the box.

The switch is held in position by two metal screws which engage in tapped lugs inside the box. Alignment of the switch is achieved either by slotted holes in the back of the box or by adjustable lugs (see Fig. 3.7).

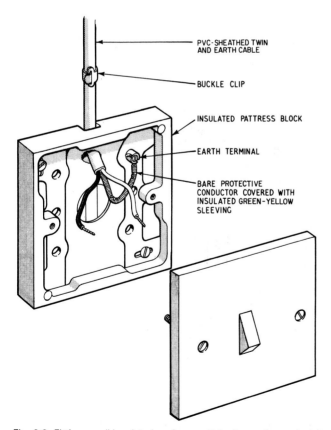

PVC-SHEATHED TWIN AND EARTH CABLE

BUCKLE CLIP

INSULATED PATTRESS BLOCK

EARTH TERMINAL

BARE PROTECTIVE CONDUCTOR COVERED WITH INSULATED GREEN-YELLOW SLEEVING

Fig. 3.8. Fitting an all-insulated surface switch. An earth terminal should be provided and connected to the protective conductor, which should be covered with an insulating sleeve

Surface switches

This type of switch is normally employed with surface wiring and may be mounted on a moulded pattress block or in a metal or insulated box with knockouts for cable entry (Fig. 3.8).

Moulded and metal boxes are available which enable the square or rectangular flush-type switches to be surface mounted. Many manufacturers make boxes which are suitable for either surface or flush mounting.

The appropriate knockouts should be removed from metal boxes and either a rubber grommet fitted for surface wiring or brass bush and collar for conduit. Moulded or metal boxes are suitable for PVC-insulated surface wiring, but an earthing terminal must be provided on the switch or box (see Fig. 3.8). Metal boxes with suitable knockouts must be used if surface conduit is employed, and an earthing terminal is not normally necessary provided the conduit is electrically continuous back to the meter position.

4 Wiring lighting points

The minimum of fixed lighting required in each room, in hallways and on stairways and landings is one point controlled by one switch.

The use to be made of each room must be considered when determining the position of lighting points and switch positions, and additional lighting may be desirable in certain rooms, such as lounge, kitchen and bedrooms. This may be taken care of by additional fixed lighting points on ceilings or walls, or by socket-outlets (see next chapter) to which portable table- or floor-lamps may be connected.

Switching from more than one point should receive consideration in rooms with two or more doors, bedrooms, passages and halls, stairways and landings. Also, the lighting of deep cupboards, lofts, cellars and back and front doors should not be overlooked.

The basic lighting circuit has been described in Chapter 1, but the importance of connecting switches to the non-earthed, or phase side of the mains must be emphasised. This will ensure that switch-feeds are the only permanently 'live' parts of the installation, and that all other parts of each circuit including lampholders are 'dead' when the switch is off. A mistake which results in a single-pole switch being connected in the neutral wire can be dangerous and means that the installation does not comply with IEE Regulations.

It is usual to arrange a number of switches and lamps in parallel on the same circuit, the switch-feeds all being connected to the phase conductor and the neutral taken direct to the lighting points or ceiling roses. How this is done in practice depends largely on the system of wiring adopted. There are two methods — looping and jointing.

Looping

Looping is the system employed almost universally in conduit installations and most sheathed wiring systems.

No joints or connections are made anywhere except at recognised termination points normal to the circuit; that is, at switch terminals and ceiling roses, outlet boxes or lampholders.

Fig. 4.1 shows the method of looping normally employed with conduit

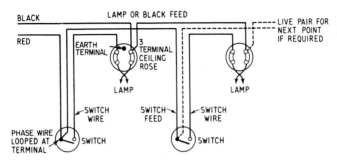

Fig. 4.1. Looping in principle of wiring a lighting circuit for use with conduit. Protective conductor not shown. On metal conduit systems this will normally be the conduit.

systems. It can be seen that much of the conduit carries three wires; this is a point to remember when estimating the amount of cable required e.g. 100 m of conduit will require approximately 300 m of cable. This does not include protective conductors, which will not normally be required in metal conduit systems.

Fig. 4.2 shows a looping-in system which can be employed with twin and earth PVC-sheathed cable using four-terminal ceiling roses. On this system the full mains voltage will exist at all ceiling roses and it is therefore essential for the phase terminal to be shrouded to prevent accidental contact.

Fig. 4.3 shows a typical layout in a small house using the looping principle with four-terminal ceiling roses. Only one circuit is shown for clarity; it has been assumed that the ground floor will be on a separate circuit to the first floor. It should be noted that the 'go' and 'return' cables must always be run in the same conduit, including wiring to two-way switches.

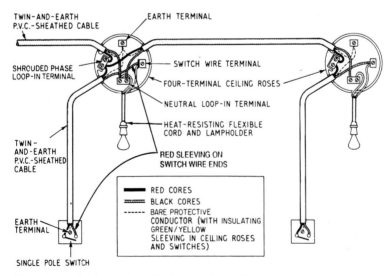

Fig. 4.2. Looping in principle with four-terminal ceiling roses for use with twin-and-earth cables

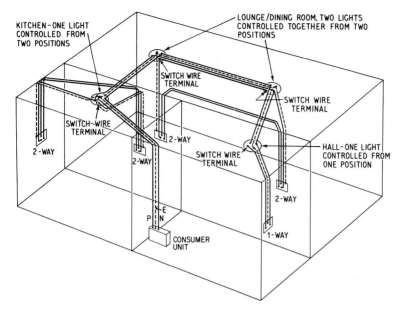

Fig. 4.3. Typical ground-floor lighting circuit using an all-insulated wiring system and four-terminal ceiling roses

An earth terminal must now be provided at all lighting points and connected to the protective conductor of the final circuit.

Jointing

The use of junction boxes with fixed brass terminals or separate porcelain- or plastic-shrouded connectors in metal or insulated boxes may be adopted, particularly for twin or twin-and-earth wiring systems (e.g., PVC-insulated and sheathed or mineral insulated). Connectors in boxes can also be used with conduit systems to save cables in looping back. The layout of lighting points using the jointing principle is illustrated in Figs. 4.4 to 4.6.

Junction boxes may give a saving in cable but, in general, all joints should

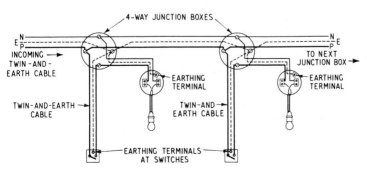

Fig. 4.4. Lighting circuit using the junction box, or jointing, principle and three-terminal ceiling roses

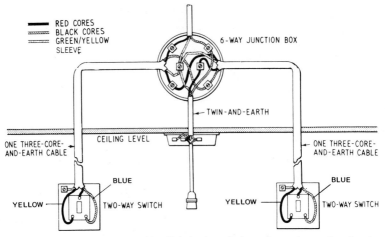

Fig. 4.5. A typical two-way switching/lighting installation using a six-way junction box. Note earthing facilities at switch and lighting points. If two twin cables are run to the two-way switches, the two cables to each switch *must* run together.

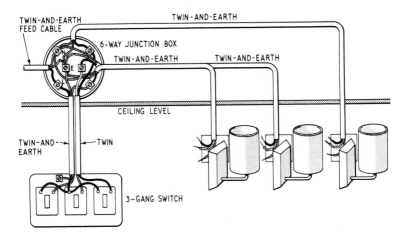

Fig. 4.6. Three wall lights can be controlled separately from switches at the doorway using a six-way junction box. Note earthing facilities at switch and lighting points

be avoided as far as possible except in special circumstances, e.g. where exceptionally long cable runs are involved. Junction boxes should be located in accessible positions.

Number of points on one circuit

It is recommended that in the case of lighting-outlets the assumed current demand should be equivalent to the connected load, with a minimum of 100 W per lampholder.

Domestic lighting circuits are almost always rated at 5 A, or 6 A, which means that if 100 W per lampholder is assumed, then up to twelve lighting-outlets could be connected to each 5 A circuit subject to a maximum of 1200 W. In practice, it is usual to divide the fixed lighting into two or more circuits, each with seven or eight lighting-outlets on it. It is also desirable to leave one or more spare ways in the consumer unit for any possible future extensions.

Cables for lighting

Lighting circuits, fed via a 5 A fuse or MCB, may be served by 1.0 or 1.5 mm² cable. 1.0 mm² cable is normally adequate but, if the circuit is a long one, it may be necessary to use a larger size to avoid excessive voltage drop—see Chapter 2.

Heat-resisting flexible cords, e.g. those insulated with butyl or ethylene propylene rubber, silicone rubber or glass fibre, are recommended for the connections between the ceiling rose and the lampholder in pendant lighting fittings where tungsten-filament lamps are to be used. Precautions are also desirable for batten lampholders in such situations, e.g. termination of the normal fixed wiring in a suitable joint box adjacent to the lampholder, the final connections being made in heat-resisting cables, or individual protection of the cable cores by sleeves of suitable heat-resisting material, e.g. silicone-bonded glass braiding. The current ratings and correction factors for various ambient temperatures for flexible cables and cords of various types are given in the IEE Regulations (see also page 49).

Two-way switch control

Two-way switching (as explained in Chapter 1) is often useful in bedrooms, halls and landings and rooms with two doors. A typical arrangement is

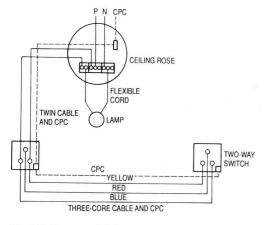

Fig. 4.7. Two-way lighting

shown in Fig. 4.7; this is an alternative to Fig. 4.5 and does not require a separate junction box.

The conversion of an existing single switch, one-way control of a lighting point to two-way control is one that is often required. It can be carried out without disturbance of the wiring already in place.

The existing one-way switch is replaced by a two-way switch, and another two-way switch is fixed at the other control point. The cable and connections are then as shown in Fig. 4.7.

Control from more than two points

Where two-way switching is installed, it may be found that one switch is not readily accessible from certain points. This is particularly applicable in long hallways, or where there are more than two doors in a room. The difficulty can be overcome by inserting a third or intermediate switch as already mentioned in Chapter 1. An intermediate switch is essentially a reversing switch without an off position, and operates by reversing the connections between the two-way switches. Conversion of an existing two-way control to three-way is shown in Fig. 4.8.

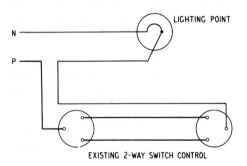

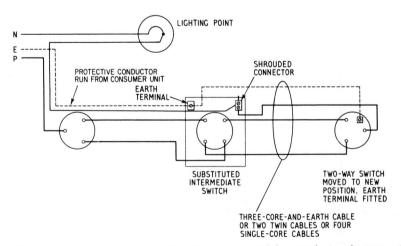

Fig. 4.8. Conversion from a two-way to three-way control. A protective conductor must be run from the consumer unit to the switches

On and off control of lighting can be extended to more than three points by the addition of further intermediate switches.

Ceiling switches

As stated in Chapter 3, a wall mounted switch for controlling a lighting point in a bathroom must not be mounted within reach of anyone using the bath or shower, and this may involve the provision of a wall-mounted switch outside the bathroom or, if control is required inside the bathroom, a ceiling switch, operated by an insulating cord, must be fitted. Two-way ceiling switches may also be used in bedrooms or kitchens, often in conjunction with a wall-mounted two-way switch. Fig. 4.9 shows a typical section of a lighting circuit supplying a bathroom and bedroom, and indicates how economies in both cable and installation costs can be effected by using ceiling switches, thus eliminating the drops to wall switches.

Ceiling switches are fitted on to a non-flammable pattress block or plate; a typical design is shown in Fig. 4.10. The pattress or plate also serves to cover any irregular holes in the ceiling and can be fitted to a standard conduit box to BS 31. An earth terminal must be included on ceiling switches, and connected to the protective conductor of the final circuit. Switch terminals will normally accept two 1.5 mm^2 cables for looping.

Dimmer switches

These can be used to control the lighting level in lounges, bedrooms etc. They usually need to be fitted in deeper boxes in lieu of a normal switch and sometimes include a local fuse and separate on/off switch.

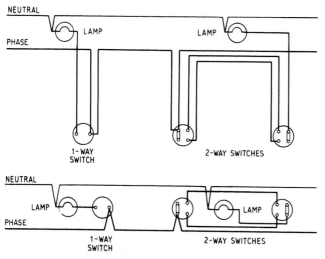

Fig. 4.9. Wiring for lighting circuit with wall switches (above) and ceiling switches (below) which allow cheaper installation

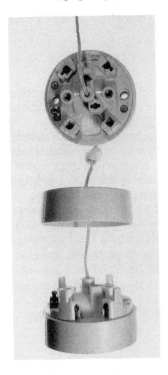

Fig. 4.10. A cord-operated 5 A ceiling switch

Lighting fittings

Where conduit systems are employed, it is usual to use a standard circular iron or steel box to BS 31, with some form of back outlet so that the lower rim of the box does not come below the surface of the plaster. This gives a convenient outlet to which a ceiling rose or ceiling plate may be fixed. The standard box has two fixing lugs drilled and tapped at 51 mm centres for fixing a direct-mounting surface or semi-recessed ceiling rose, a batten lampholder or, where connectors are used in the box, a pendant or hook plate. Wall bracket lights may be similarly mounted.

To conceal the joint between the box and plaster, a moulded break joint ring may be interposed between the box and the ceiling rose or plate.

Looping-in boxes are employed in concrete floors; these may have two, three or four 16 or 20 mm clearance holes in the back of the box to enable conduits to be attached by means of lock-nuts and brass-ring bushes.

Where non-sheathed cables are used in conduit, or where sheathed cables have the sheathing removed for connection to a lighting fitting, then that fitting must be mounted so that the non-sheathed portion of cable is enclosed in a non-flammable box. This is achieved by mounting the fitting on a pattress block, or plate, or by installing a fitting which has a fully insulated integral base.

Ceiling roses and lampholders

A ceiling rose is used to make a safe and efficient connection between the circuit wiring of an installation and the flexible wires of pendant lights fed from the wiring.

Most modern ceiling roses are of the moulded-plastic type with an integral mounting block. This means that unsheathed cables can be enclosed within the fitting and there are no live metal parts exposed at the rear face.

The IEE Regulations require that an earth terminal be provided at all lighting points and connected to the protective conductor of the final circuit. Also, if a looping-in terminal is used, so that it remains 'live' when the associated switch is off, it must be shrouded so that it cannot be accidentally touched when the rose is dismantled to the extent necessary to replace the flexible cord. These features are illustrated in Fig. 4.11, which shows a four-terminal rose for surface mounting.

This type of ceiling rose has a moulded plastic backplate with the terminals moulded into it. The three-terminal type has two triple-hole terminal blocks for incoming phase and neutral wires, a double-hole terminal for the switch wire and a terminal for the CPCs.

The two flexible cord wires from the lampholder are taken under the cordgrip lugs before connection to the terminal blocks. Some kind of cord grip must be fitted to take the weight of the lampholder or fitting so that there is no strain on the connections at the terminal. The following table gives

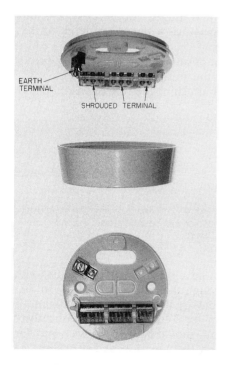

EARTH TERMINAL

SHROUDED TERMINAL

Fig. 4.11. Ceiling rose showing looping-in terminals which are shielded to prevent accidental contact. An earth terminal is fitted into the backplate.

the weight to which a flexible cord may be subjected when it supports, or partly supports, a lighting fitting as follows:

Cross-section of conductor (mm²)	Maximum weight (kg)
0.5	2
0.75	3
1.0	5

The type of outlet-box used also has a bearing on this, since a thermoplastic (e.g. PVC) type of box, if used for the suspension of a lighting fitting, must not support more than 3kg. This obviously applies both to ceiling roses with integral boxes and semi-recessed ceiling roses attached to plastic ceiling boxes.

The ceiling rose cover should be slipped over the flexible cable before the cable is connected to the rose, and fixed to the rose after the connections have been made. Some covers screw on and others have two fixing screws.

Fixing and connecting a ceiling rose

When the position of the lighting-outlet has been decided, a hole is made in the ceiling at the appropriate point, if possible, adjacent to a joist so that a good fixing for the fitting can be obtained. Often with new housing, the wiring is carried out prior to plastering and is clipped along a convenient joist and approximately 100 mm is left hanging below the ceiling level, so that plastering is carried out round the wires.

If it is not possible to arrange the light position adjacent to a joist, then a bearer-board must be fixed between the joists as shown in Fig. 4.12. Longer wood screws should be used for attaching the rose to ensure that a good fixing into the board is obtained.

Fig. 4.12. Fixing a ceiling rose in a position which lies between joists using a bearer board

When the lamp feed and switch wires have been pulled in, together with any looping wires, they are passed through the hole in the ceiling so that approximately 50-100 mm 'slack' cable is showing. The insulation is removed from the ends of the cables for 12-20 mm and, if sheathed twin-and-earth cables are used, the sheath should be stripped back only far enough to enable connection of the wires without tension, and all unsheathed portions of cable must be inside the block, box or integral fitting. Any looping cables should be carefully twisted together, and single wires bent over double to ensure a good connection.

If a break ring or pattress block is to be used, it is passed over the cables at this stage, any knockouts being carefully removed and rough edges smoothed down. Deep pattress blocks, with tapped fixings for the rose, are screwed directly to the joist or bearer board, through the ceiling plaster or ceiling board.

With the integral-base type of rose, cables are passed through the back knockout and the bare ends pushed through the appropriate terminal sleeves or blocks so that approximately 3 mm protrudes, and screws tightened. Care must be taken to ensure that the looping-in wires are connected to the shrouded terminal on three-terminal roses, and also that if flat twin-and-earth cables are used there is no danger of the bare protective conductor coming into contact with the other terminals. This can be ensured by passing a piece of green and yellow coloured plastic sleeving over the protective conductor before connection and red over the switch wires.

When the terminal screws have been tightened, the slack wire is pushed back through the knockout, making sure that all unsheathed sections remain inside. The base is then screwed through the ceiling to the joist or bearer-board with countersunk woodscrews.

Connecting a light pendant to a ceiling rose

Heat-resisting flexible cords should be used for connecting the lampholder to the ceiling rose (see 'Cables for Lighting' page 43).

The sheathing on the flexible cord is cut back for approximately 40 mm and the insulation removed for approximately 10 mm. The two cores of the cord are then passed round or through, or arranged to be held by, the cord-grip device in the ceiling rose and the bared ends connected into the two appropriate terminal blocks. Care must be taken to ensure that the flexible cord is connected to the neutral and switch-wire terminals.

Certain types of lampholders and fittings require to be earthed, and in this case a three-core flexible cable must be used, the green and yellow coloured core being used as the protective conductor.

The ceiling rose cover is passed over the flexible cord and fixed into place — care being taken with cord-grip devices which rely on pressure from the cover to ensure that cables secured are positioned correctly. Not more than one outgoing flexible twin cord must be connected to a ceiling rose, unless it is specially designed for multiple pendants.

Bayonet-type lampholders

The connection to the lamps of an installation from the flexible wires or circuit cables is made with a lampholder, holders of the bayonet type being normally used for lamps whose rating does not exceed 150 W. Most bayonet lampholders have retracting contacts of either of two main types, solid-plunger or spring-plunger (see Fig. 4.13).

The solid-plunger type is of one-piece construction, having an external spring, and with the wiring terminal integral with one end of the body, through which current flows to the lamp. To ensure unobstructed movement of the plunger and avoid breakage of the wires, only flexible cords should be used for wiring this type.

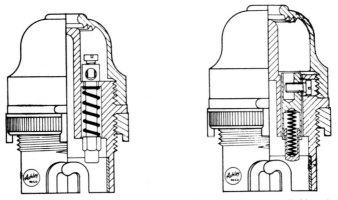

Fig. 4.13. Two types of bayonet lampholder: a solid plunger (left) and a spring plunger (right)

The spring-plunger type is of two-part construction, with the contact part retracting against a spring inside a fixed barrel with integral wiring terminal, through which current flows to the lamp. As the terminal remains stationary during plunger contact movement, suitable cables as well as flexible cords may be used for wiring.

IEE Regulation 601-11-01 requires that in a room containing a fixed bath or shower cubicle, parts of a lampholder within 2.5 m from the bath or shower cubicle must be constructed of, or shrouded in, insulating material and, in addition, bayonet-type lampholders must be fitted with a protective shield complying with BS 5042. This condition is normally met by fitting a screw-on plastic 'skirt'. Alternatively, totally enclosed lighting fittings may be used. Whenever possible, a lighting fitting should be out of reach of a person using a fixed bath or shower.

The above precautions should also be taken in any damp situation, or in any situation where the lampholder can be touched by a person in contact with, or standing on, an earthed object.

Lampholders constructed from metal must be earthed.

All bayonet lampholders B15 and B22 must be of the type T2 temperature rating.

Cord grip of pendant lampholder

Pendant lampholders are fitted with a cord grip which takes the weight of the lampholder, lamp and shade and avoids strain on the connections between the flexible wires and the plunger terminals. Most lampholders have a cutaway section on opposite sides of the centre insulating bridge so that flexible cords are brought down either side of the bridge, under the cutaway section and connected to the terminals from opposite directions.

Lampholders with a special gland-type grip for use with round PVC-sheathed flexible cables are also available. Whatever type of cord grip is incorporated, it should never be omitted.

Batten lampholders

This type of holder is used for lighting points fixed directly to ceilings or walls. Ceiling-mounted batten holders can be used in bathrooms, washhouses and porches. Wall-mounted, 45°-angled batten holders are especially suitable over bathroom mirrors and kitchen sinks. To comply with IEE Regulations batten-type lampholders used in bathrooms should be fitted with a screw-on plastic skirt.

An earth terminal must be provided at all batten-type fittings; this may be an integral part of the batten fitting or on the pattress block, if one is used. Pattress blocks with four terminals (earth, phase loop-in, switch wire and neutral) can be used with some batten fittings; this enables the connection to the batten lampholder to be made with heat-resisting cables.

Some types of moulded plastic batten-type lampholders are suitable for use without a plastic-mounting block for direct attachment to standard circular iron boxes.

Batten lampholders can be used on either concealed or surface wiring systems, as the moulded back is provided with an outlet at the back and knockouts for cable entry.

Edison-screw lampholders

With this type of lampholder the lamp cap is provided with a coarse thread which is screwed into the lampholder. This connection makes one 'pole' of the circuit. A central metal stud in the lampholder makes contact with a central terminal plate in the top of the lamp cap; this connection makes the other 'pole'. The phase wire must be connected to the centre terminal of the lampholder, and the neutral to the threaded section (IEE Regulation 553-03-04).

Edison-screw lampholders are available in cord grip, screwed or batten types, and must be shrouded in insulating material and be drip-proof.

Adaptors and switched lampholders

Lampholder adaptors may be used to make temporary connection to small-current portable electric appliances from a lampholder, **but this practice is not recommended**. The full rated current of the portable appliance should not exceed 1 A. Lampholder adaptors should never be used when the floor of the room in which the portable appliances are used is hygroscopic, or in bathrooms, basements, lavatories, washhouses, etc.

A switch may be incorporated with a lampholder. The switch is operated by a push bar which passes through the case of the lampholder. Such switched lampholders are often fitted to standard and table lamps. However, they must not take the place of a wall switch or socket-outlet to control the circuit completely, and must not be used in bathrooms.

Fluorescent-tube lighting

The three most important features of low-pressure mercury fluorescent-tube lighting may be summarised as follows:

1   Lighting efficiency is very much higher than that of metal-filament lamps. For example, one 85 W 'natural' fluorescent lamp gives almost as much light as four 100 W metal-filament lamps; and one 85 W 'warm white' lamp gives more light output than five 100 W metal-filament lamps.
2   The life of a fluorescent lamp is approximately seven times that of a metal-filament lamp.
3   Light is spread over the surface of the tube so that there is a much more even distribution of light and absence of glare which has to be guarded against when the older types of lamps are used.

There are several types of colour output from fluorescent lamps, but probably the most suitable for use in the home are 'natural', 'warm white', or 'deluxe warm white'.

Except perhaps in a bathroom, kitchen, garage or workshop, fluorescent tubes will generally be located out of the line of sight behind baffles, pelmets or cornices, or screened behind diffusers. Their high lighting efficiency and linear distribution of light give them advantages over metal-filament lamps for applications such as the lighting of areas of walls, ceiling, curtains, niches and pictures, false windows, luminous shelves, luminous beams, etc.

Principles of the fluorescent tube

A typical low pressure mercury fluorescent lamp consists of a glass tube, 38 mm in diameter and between about 600 mm and 2400 mm long. The tube is filled with argon or other gas at a pressure of between $2\frac{1}{4}$ and 3 mm of mercury (approximately 1/250 of atmosphere pressure) and also contains a drop of liquid mercury. The interior surface of the tube is coated with a fluorescent powder, the phosphor, which converts the ultra-violet light produced by the discharge into visible light.

At each end of the tube are electrodes, which serve the dual purpose of cathode and anode, since these lamps are used in a.c. circuits, where the current flow reverses every half-cycle. The cathodes of a hot-cathode fluorescent lamp consist of a coiled coil, or braided tungsten, filament coated with a barium-oxide thermionic emitter and held by nickel support wires. Anodes in the form of metal strip or wires attached to the support wires, are present in the tubes of high loading, but in the tubes of lower loading the support wires themselves act as anodes. The inert gas is used to reduce the rate of evaporation of barium from the cathode and also to reduce the starting voltage.

The tube cathodes are heated by passing a current through each filament, which causes a large number of electrons to be emitted by the oxide coatings. Assuming sufficient voltage is applied across the tube, a glow discharge is set up through the argon gas which excites or ionises the mercury atoms throughout the tube and establishes a mercury-arc discharge.

All fluorescent tubes require appropriate control gear for operation: a ballast to limit current through the tube, a starting device for preheating the cathode filament and, usually, a power-factor-correcting capacitor. The circuit is stabilised by inserting a current-limiting ballast, usually of the

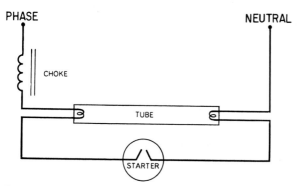

Fig. 4.14. Basic circuit for fluorescent lighting

choke-type, in series with the tube. As the arc current increases, the voltage drop across the ballast increases, and the voltage across the tube decreases, until the point of equilibrium is reached (see Fig 4.14).

The simplest starting device, and the most used, is the starter switch. When the starter switch is closed, a current flows through the ballast and the two cathodes in series and, as soon as the cathodes reach emission temperature, local ionisation is set up and the ends of the tube begin to glow. The starter switch is then quickly opened, and sufficient voltage is applied across the tube to cause the arc to strike. This is due to the sudden change of current and the self-inductance of the iron-cored choke.

Starting switches

Automatic starting switches allow a heating current to flow for a predetermined time before they open and produce the striking pulse. There are two distinct types of starting switch, namely: the glow type and the thermal type.

In the glowswitch type of automatic starter, the switch contacts are mounted on bi-metallic strips which bend towards each other when heated. The contacts are sealed in an argon gas-filled bulb, and the whole assembly, together with the radio-interference-suppression (r.i.s.) capacitor is housed inside a canister (see Fig. 4.15).

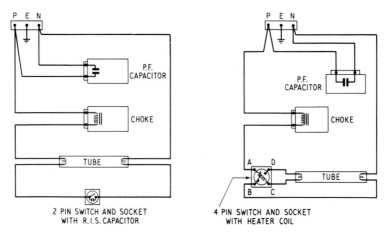

Fig. 4.15. A glowswitch automatic starter circuit (left) and a thermal switch starter circuit (right)

When the circuit switch is closed, full mains voltage is applied across the starter switch, and this causes a glow discharge to be set up across the open contacts. The discharge heats the bi-metallic strips and thus causes the switch to close. As soon as the switch closes, current is applied to the cathodes which produces the pre-heating mentioned earlier. The discharge is also extinguished and the bi-metallic strips start to cool down. After a predetermined time the contacts spring apart and the voltage surge is induced in the choke and applied across the tube, causing it to strike.

These starters are now being superseded by electronic, solid state starters which are claimed to have a longer life, require less maintenance, and cause less wear on the lamps.

The thermal-type starter circuit is now little used, but a description is given for reference purposes. The switch contacts are mounted on bi-metallic strips which are connected to terminals B and D, but a small heater coil is mounted close to the strips and connected to terminals A and C (see Fig. 4.15). When the circuit is closed, the starter-switch contacts are closed and current flows through the choke, heater coil, tube cathode, starter switch and the other tube cathode as shown. The heat from the heater coil causes the contacts to spring apart after a predetermined time and cause the necessary voltage surge to be induced.

Power-factor capacitor

The presence of the iron-cored choke, being an inductance, means that the circuit power factor could be as low as 0.5 lagging. This would mean that a large wattless current would flow in the main-feeder cables and it is necessary, therefore, to improve the power factor to about 0.85 lagging. This is done by connecting a suitably rated capacitor, often referred to as a 'p.f.c.' or 'shunt' capacitor, across the input terminals of the fitting.

Twin-tube and quickstarting circuits

Two tubes may be mounted in a single fitting, each tube normally having its own starting switch. Only one choke and p.f.c. are normally fitted, and the connection details are given in Fig 4.16.

A fluorescent-tube circuit designed to give a quick- or switchless-start is shown in Fig. 4.17. With the circuits mentioned previously, the use of a starter switch means that several seconds may elapse between switching on and the tube striking. With quickstart circuits, this starting period is reduced considerably by arranging that the discharge strikes automatically as soon as the cathodes reach emission temperature.

Switchless-start circuits take two main forms: the transformer circuit, in which the cathodes are heated from a transformer; and the semi-resonant circuit, in which the lamp cathodes are connected in series via a choke

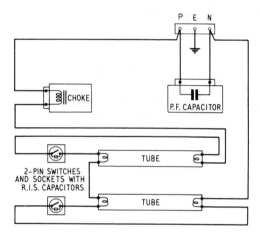

Fig. 4.16. A typical twin-tube circuit with two glowswitches

winding and a capacitor, with part of the choke winding acting as a current limiter after the discharge is established.

A typical transformer unit comprises an iron-cored auto-transformer with secondaries at each end which provide 10-12 V across each tube cathode when 200-250 V is applied to the primary.

Universal or GP fluorescent tubes are suitable for use on either switch-starter or starterless circuits. On starterless circuits, an electrical charge on the lamp is sufficient to repel electrons and thus prevent starting. Universal tubes are of two types: MCFE/U, which is coated externally with silicone, a moisture repellant; and type MCFA/U, which is provided with a metallic strip affixed externally along the lamp. With the silicone-coated tube, provided that an earthed metal strip is mounted close to the tube (within 12 mm) along its entire length, then the tube will strike automatically. With most fittings, the earthed metalwork is provided by the fitting itself. Where tubes are installed without metal fittings, as in indirect-lighting applications,

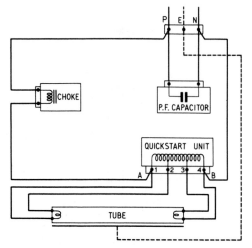

Fig. 4.17. A quickstarting circuit. For voltages below 220 V, leads A and B are connected to terminals 2 and 3

a 50 mm wide earthed metal strip can be fixed behind the length of the lamp, or the tube must be of the type that has a metal strip cemented to the outside of the tube; this strip must be connected to both lamp caps and earthed by means of a brass lampholder (bayonet-cap tubes) or special clips (tubes with bi-pin lampcaps).

Once the arc has been established, the voltage across the tube (and also across the quickstart unit) falls to approximately half mains voltage. Auxiliary cathode heating is therefore reduced automatically.

Semi-resonant-start circuits

Semi-resonant-start circuits have been developed for some tubes, e.g. 1500 mm (65 W) and 1800 mm (85 W) tubes. The mains voltage (240 V) is applied to the circuit and the pre-start current flows from the line terminal through the primary winding, the right-hand tube cathode, the secondary winding, the p.f. capacitor and finally through the left-hand tube cathode and back to the neutral (see Fig. 4.18). The r.i.s. capacitor is for radio interference suppression.

The pre-start current quickly heats up the cathodes until emission begins and, as the circuit is mainly capacitive, the pre-start current leads the mains voltage. The primary and secondary windings of the ballast are wound on the same laminated-iron core in opposition, which means that the primary and secondary voltages are 180° out of phase. The mains voltage of 240 V is increased to about 280 V across the tube by resonance effect, i.e. the secondary voltage of the ballast is added to the capacitor voltage. This high pre-start voltage across the tube gives reliable starting, even in cold environments.

As with quickstarting circuits of the transformer type, earthed metalwork must be provided adjacent to the tube for satisfactory operation. Once the

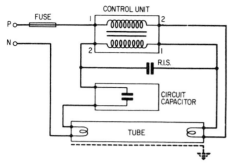

Fig. 4.18. A semi-resonant start circuit. The earthed metalwork must be within 12 mm of the tube

tube arc has struck, the current in the primary winding builds up in the same way as a choke, and the voltage across the primary winding is about 190 V when the tube current is stabilised.

To protect the semi-resonant-start circuit against overheating of the ballast windings due to capacitor failure, an external h.b.c. cartridge fuse (usually 1A) is provided in the supply terminal block or in a separate fuse-holder on each fitting.

Circular fluorescent fittings

Circular tubes are also manufactured in various diameters, and a special control fitting of either the starter-switch or transformer type can be used with them. The 400 mm diameter circular tube is rated at 40 or 60 W and provides a compact light source for flush- or pendant-ceiling mounting. A standard circuit diagram is given in Fig. 4.19.

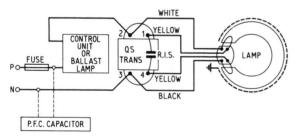

Fig. 4.19. A circular fluorescent tube circuit. The control unit comprises either a choke and p.f.c. or filament ballast lamp (*Thorn Lighting*)

Resistance ballast circuits

Fluorescent tubes can be operated in series with a resistance which takes the place of the normal choke current-limiting ballast and p.f. capacitor. As there is no choke to provide a voltage impulse, a siliconed or metal-strip tube should be used. The ballast resistance may be in the form of a special tungsten-filament lamp, or a wire resistor. Fig. 4.20 shows the circuit for a

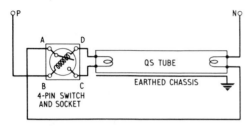

Fig. 4.20. A filament ballast lamp used with a glowswitch starter

fitting incorporating a special combined starter-ballast lamp; the filament ballast lamp is cemented into the end of a standard four-pin starter canister. The ballast lamp is connected across terminals D and B whilst the glowswitch starter is across A and C. The filament ballast lamp therefore takes the place of the normal choke and p.f. capacitor.

In situations where no earth is available, an artificial-earth unit is often connected between the neutral lead and the metal chassis to assist the starting of the tube. The artificial-earth unit consists of a suitable resistor and capacitor connected between the available metalwork and mains neutral.

Series operation and d.c. operation

When used on 200-250 V mains, lamps of 600 mm and 450 mm should be connected in pairs of the same wattage rating in series, not singly. Only one suitable choke and p.f. capacitor is required per pair of lamps; a single starter-less unit can be used to start both lamps, while a starter circuit requires a starter switch connected across each lamp.

Satisfactory operation of fluorescent lamps can be achieved on d.c. supplies, but the power consumed is about double that on a.c., owing to the series resistance (of the correct value and rating) which must be connected in series with each choke. A polarity-reversing switch should be fitted in the mains, so that it can be operated, as required, to avoid tube darkening at one end due to mercury migration. The p.f. capacitor is omitted. A d.c. glow or thermal-type starter should be used. However d.c. operation is not now very common. It is more usual to convert the d.c. supply to a.c. by means of an 'inverter' within the lighting fitting; standard a.c. circuits can then be used. This is often done on fittings which are to operate on an emergency battery supply, which is, of course, d.c.

Fluorescent-tube circuit wiring

The basic circuit-wiring requirements are essentially the same as for ordinary filament lamps, but there are a number of specific points to consider.

It is recommended that a switch not specifically designed to break an inductive load of its full rated capacity in accordance with BS 3676 shall, if used to control a discharge-lamp circuit, have a current rating not less than twice the total steady current which it is required to carry. Many modern switches meet this requirement and do not require to be derated. In case of doubt, the manufacturer should be consulted.

In considering the current taken by a fluorescent fitting, it is not sufficient merely to convert the nominal lamp watts to the equivalent current. Additional current is taken owing to losses in the control gear, harmonics and power factor, and these values depend on the circuit used. If the particular fitting to be used is known, the actual current it takes can be obtained from manufacturer's data. However, when more precise information is not available, the demand on discharge lighting circuits in voltamperes should be taken as the rated lamp watts multiplied by not less than 1.8; this takes into account control gear losses and harmonics and assumes that the power factor is not less than 0.85 lagging.

However appreciable numbers of fluorescent fittings are not likely to be encountered in domestic installation.

5 Wiring socket-outlets and portable appliances

Socket-outlets provide an easy and convenient method of connecting portable electrical apparatus to the supply. The socket-outlet is permanently connected to the circuit cables and has shaped entries which allow only the correct type of plug to be inserted, the plug being connected to the portable appliance via a flexible cord.

IEE Regulation 533-01-07 requires that 'provision shall be made so that every portable appliance and portable lighting fitting can be fed from an adjacent and conveniently accessible socket-outlet.' It should be assumed that the length of flexible cord on portable appliances will be 1.5 to 2 m.

Account must be taken of the fact that in various rooms, particularly living-rooms and kitchens, more than one appliance will often be in use simultaneously. This must be allowed for, so as to avoid the use of inconvenient, and possibly unsafe, socket-outlet adaptors and long flexible cords. With new buildings, the emphasis should be on anticipating the future electrical needs, thus avoiding the possibility of an unqualified householder attempting to extend the installation himself.

All socket-outlets should be fitted well above floor level or working surfaces. A minimum clearance of 150 mm is recommended. Switched socket-outlets are not essential but are recommended for most locations so as to avoid frequent removal of plugs and consequent arcing and wear on contacts.

Table 5.1 gives the recommended provision for socket-outlets in domestic accommodation. This is intended primarily for local authority housing and it may be desired to increase the quantities in some cases. It is worth noting that a twin socket costs very little more than a single one if provided on the initial installation.

Outlets for cleaning appliances should be included in halls and passages and on landings as necessary.

The following general recommendations can be applied to semi-detached or detached houses and bungalows; discretion must be used with smaller properties.

Living rooms Socket-outlets should be provided on each side of the fireplace

Table 5.1. Recommended socket-outlet provisions

Part of dwelling	Desirable provision	Minimum provision
Working area of kitchen	4	4
Dining area	2	1
Living area	5	3
First (or only) double bedroom	3	2
Other double bedrooms	2	2
Single bedrooms	2	2
Hall and landing	1	1
Store/workshop/garage	1	–
	20	15
Single study-bedrooms	2	2
Single bedsitting rooms in family dwelling	3	3
Single bedsitting rooms in self-contained bedsitting room dwellings	5	5

These figures refer to individual outlets. A twin socket counts as two outlets.

and others on the opposite side of the room where they are least likely to be masked by furniture. A twin socket should be provided at the television position.

Kitchens Socket-outlets should be either at low level or above working surfaces according to their intended use. They should not be fitted adjacent to sinks unless unavoidable. A cooker should preferably be fed via a simple 30 A or 45 A switch on a separate circuit, and an adequate number of socket-outlets provided elsewhere. A socket-outlet on a cooker control unit can lead to danger due to flexible cables trailing over the top of the cooker.

Bedrooms Socket-outlets should preferably be provided at each side of the bed with other sockets as required. When fixed wall mounted heaters are installed, bedside switching may be adopted.

Shaver units Socket-outlets must not be installed in a room containing a fixed bath or shower, except that shaver supply units complying with BS 3052 may be installed. These units provide an earth-free supply via an isolating transformer and include protection against overload. On some units the output is 240 V only whilst others provide outputs of 240 and 115 V.

For locations other than bathrooms or showers, simpler shaver supply units are available and these must comply with BS 4573; protection against overload is again included.

All shaver supply units should be labelled *For shavers only*.

Types of plugs and sockets

Socket-outlets can be obtained in 15 A, 13 A, 5 A and 2 A ratings. They can be surface- or flush-mounted and switched or unswitched. However, the 13 A rectangular section three-pin, non-reversible socket-outlet with its accompanying 13 A fused plug is the standard fitting employed for new installations in domestic premises. The other types of sockets and plugs are now obsolete, but as they are still in use and replacement parts are still available, they are briefly covered in this chapter.

Two-pin plugs are not recommended for connection of any apparatus except electric clocks and shavers. Special socket-outlets designed for electric clocks and shavers can be obtained; both must be protected by a 3A fuse or other equivalent current-limiting device.

All plugs and socket-outlets must conform to the appropriate British Standard as given in Table 5.2.

Table 5.2. Plugs and socket-outlets for low-voltage circuits

Type of plug and socket-outlet	Rating (A)	Applicable British Standard
Fused plugs and shuttered socket-outlets, 2-pole and earth, for a.c.	13	1363 (fuses to 1362)
Plugs (fused or non-fused) and socket-outlets, 2-pole and earth	2, 5, 15, 30	546 (fuses, if any, to 646)
Plugs (fused or non-fused) and socket-outlets, protected type, 2-pole with earthing contacts	5, 15, 30	196

Socket-outlets and plugs made to British Standard Specifications incorporate a number of important safety features. During normal insertion or withdrawal of the plug, it is not possible for the fingers to touch live metal. It must not be possible to insert the plug in the socket so that a pin overhangs the outlet. The spacing of the pins and their lengths and diameters are specified, and the pins are fixed so as to prevent them from working loose. The contacts in the socket are self-adjusting as to pitch and contact-making. There is only one hole or gland in the plug for the entry of the flexible cord and there is an effective cord grip. The cord enters by the side and not from the centre of the plug, the object of this being to discourage attempts to withdraw the plug by pulling the flex.

Connections to socket-outlets and plugs

The terminals in a socket-outlet must be connected as indicated in Fig. 5.1, where E represents the earthing terminal, L the terminal for the phase conductor and N the terminal for the neutral conductor (viewed from the

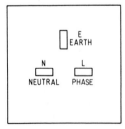

Fig. 5.1. Connections to a socket-outlet viewed from the front

front of the socket). To avoid making mistakes in wiring socket-outlets, BS Specifications require that the sockets shall be marked as indicated.

Plugs must also be marked with the letters L, N and E or with appropriate colours (see page 68) to indicate the phase (formerly 'live'), neutral and earth pins respectively when connecting flexible cords, the cores of which are similarly coloured.

Exposed metalwork of all apparatus (except appliances which are double-insulated) must be connected to the appropriate protective conductors. This is effected on portable apparatus through the protective conductor of the flexible cable and the earthing pin or earth-contact of the plug. The earthing socket of a socket-outlet is connected to the protective conductor of the final circuit, thus providing a satisfactory and continuous earth path, from the portable apparatus to the main earthing terminal.

With metal-sheathed wiring systems (including metal conduit, ducts, trunking or metal-sheathed cables), which employ the metal as the protective conductor and do not have a separate protective conductor, special earthing provisions at socket-outlets must be made.

IEE Regulation 543-02-07 requires the earthing terminals of socket-outlets to be connected to the protective conductor of the final circuit. An earthing terminal must therefore be attached to the box, trunking or other metal enclosure and connected by a separate conductor to the earthing terminal of the socket-outlet. This bonding connection must be made regardless of type of lugs employed on the box, as it makes sure that screws and fixing bolts are not relied on for earth continuity. (See Chapter 10, 'Safe and Efficient Earthing'.) The earthing terminal may also be attached to moulded insulated socket-outlet boxes and the bonding connection made as above; this is of assistance if the socket-outlet concerned is being used for looping to a non-fused spur.

The protective conductor must be insulated and coloured green and yellow except where it is constructed of copper strip or contained in a composite cable, where it may be uninsulated. Where composite cables are employed with an uninsulated protective conductor, at terminations or positions where the sheath has been stripped off the protective conductor must be enclosed in insulating sleeving coloured green and yellow.

13 A standard socket-outlet and circuits

The 13 A standard all-purpose socket-outlet with fused plug constitutes a universal size for domestic installations. The socket is designed so that appliances of any rating up to 13 A (3 kW on 240 V) can be plugged into it, the flexible cable of the appliance being fitted with the standard fused plug. Sockets and plug pins are of distinctive rectangular section for the purpose of preventing connection of any other type of plug to the socket and ensuring thereby that all flexibles are always protected by local fuses. Shuttered sockets are employed to prevent insertion of small objects.

A fuse link of the cartridge type is used in the body of the plug and is mounted between the line terminal and the corresponding plug pin. It is essential to use a fuse of the rating appropriate to the appliance to which the plug is connected. The fuse ratings available are detailed in Chapter 1 (page 3), and Table 2.1 (page 13) will assist in determining the fuse ratings required for various loadings.

The fact that all flexibles and equipment connected to 13 A socket-outlets are protected by local plug fuses affects the wiring of these socket-outlets in three ways:

1   The protective device for the circuit feeding the socket-outlets need only be selected with reference to the rating of the fixed wiring.

2   Advantage can be taken of the diversity factor available in connection with socket-outlets — that is, the probability that not all the socket-outlets will be operating at their maximum rating at the same time; this allows a number of socket-outlets, with total rating in excess of the rating of the circuit, to be looped in on one circuit (see Fig. 5.2).

3   A ring-circuit can be employed.

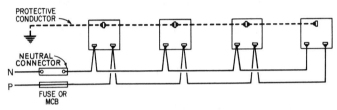

Fig. 5.2. Radial circuit for standard 13 A socket-outlets with fused plugs. Wiring rules applying to radial circuits are given in Table 5.3.

There are several possible arrangements for circuits feeding BS 1363 socket-outlets and these are summarised in Table 5.3. Note that for the purpose of this table, each twin socket-outlet must be counted as two outlets. The actual connection of stationary appliances to radial or ring circuits is dealt with in Chapter 6.

Table 5.3. Final circuits using BS 1363 socket-outlets

Type of circuit	Over-current protective device		Minimum conductor size. PVC or rubber-insulated cable (mm^2)	Maximum floor area served (mm^2)
	Rating (A)	Type		
Ring	30 or 32	Any	2.5	100
Radial	30 or 32	Cartridge fuse or MCB	4	50
Radial	20	Any	2.5	20

The ring final circuit

The ring final circuit is formed, as its name suggests, by wiring the phase, neutral and protective conductors of a power circuit in a complete ring. The two ends of the phase conductor are brought back to the same fuse or MCB on the consumer unit, and the two ends of the neutral and protective conductors brought back to the appropriate terminal blocks in the consumer unit.

Only 13 A socket-outlets conforming to BS 1363 may be used on a ring circuit, in conjunction with 2.5 mm^2 (minimum) cable (assuming PVC or rubber-insulated) and a 30 A fuse or MCB at the consumer unit. The normal rating of the cable is around 20 A, but the fuse or MCB must be of 30 A rating, since the overall current in the circuit may well be of this order. The cables are not overloaded since the current splits both ways round the circuit, and if the cable runs and socket positions are carefully planned, an even distribution of current can be obtained.

Ring and radial circuits are permitted to supply an unlimited number of socket-outlets, subject to the maximum floor areas stated for domestic premises, and also provided the estimated maximum demand on the circuit does not exceed the rating of the overload protective device. For domestic installations special consideration should be given to the loading on socket-outlets in kitchens, which may warrant a separate circuit. In family dwellings (Fig. 5.3) it is normally desirable to provide at least two ring circuits.

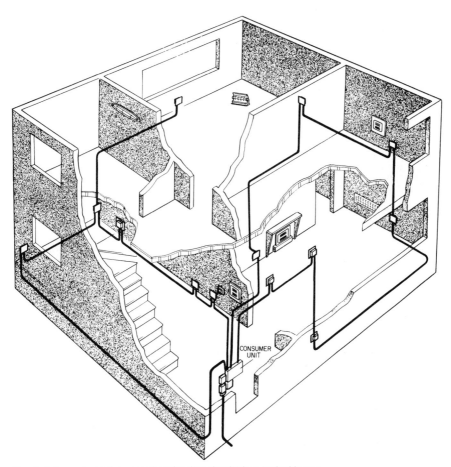

CONSUMER UNIT

Fig. 5.3. Layout of 13 A socket-outlet ring circuits in a typical house

Each ring circuit must be terminated at its own 30 A fuse or MCB, and the socket-outlets and fixed appliances should be reasonably distributed among the rings.

Spurs may be connected to ring circuits. The total number of spurs is unlimited but the number of non-fused spurs must not exceed the total number of socket-outlets and fixed appliances connected directly in the ring. A *non-fused* spur must be connected either at the terminals of a socket-outlet, at a joint box, or at the origin of the ring in the consumer unit. It must have a current rating not less than that of the conductors forming the ring and must have only one single or one twin socket-outlet, or one stationary appliance connected to it.

A *fused* spur must be connected to the ring circuit through a fused connection unit, the rating of the fuse in the unit not exceeding that of the cable forming the spur, and in any event not exceeding 13 A. When a fused spur serves socket-outlets the minimum conductor size for PVC or rubber-insulated cable is 1.5 mm².

Fused connection units for ring final circuits

Permanently connected appliances (i.e. not connected via a plug and socket) on a ring or radial circuit must be locally protected by either (a) a fuse of rating not exceeded 13 A and controlled by a switch, or (b) a circuit-breaker of rating not exceeding 16 A.

Permanently connected appliances are normally connected through fused connection units. The switch required by the above regulation may be incorporated in the connection unit or may be mounted on the appliance, provided that the connections are so arranged that the appliance can be dismantled for maintenance without exposing any parts which remain 'live' when the switch is open. Switched fused connection units are usually of the double-pole switching type, this type being suitable for all purposes. Fig. 5.4 illustrates three methods of connecting a stationary appliance to a ring circuit using a fused connection unit.

Appliances normally connected via a fused connection unit include: wall-mounted or inset fires and heaters, fixed extractor and cooling fans, hand driers and isolated lighting circuits.

Although it is permissible for a fused spur to feed 13 A socket-outlets, such cases are not very common and a non-fused spur is normally used. However a fused spur may give a saving in cable costs if exceptionally long runs are involved.

Immersion heaters fitted to storage vessels of more than 15 litres capacity, or permanently connected heating appliances forming part of a comprehensive heating system, should be on separate circuits and not connected to ring circuits. In fact it is good practice to connect any water heater with a rating of more than 1.5 kW on a separate circuit.

Fused connection units are available in many different forms, both insulated and metalclad, flush and surface, switched and unswitched. All types can be obtained with a flex outlet on the front for connection of fixed appliances by a flexible cord; where flex outlet fused connection units are

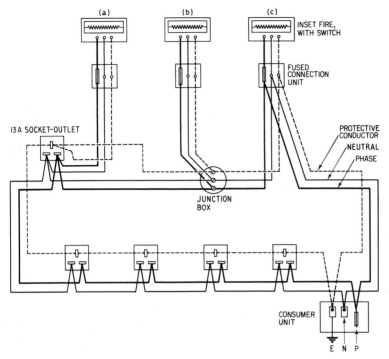

Fig. 5.4. Three methods of connecting a fixed appliance to a ring circuit using fused connection units: (a) looping from a socket-outlet; (b) wiring from a junction box; (c) looping wiring direct to the connection unit

used, the cord-grip device provided must always be tightened after connection of the flexible cable cores.

Fused connection units feeding fixed appliances, having no visible indication that they are on, should include an indicator light — normally a small neon.

Flush-type fused connection units may be fitted to standard BS 1363 or BS 3676 boxes. Pattresses are also available to convert flush-type to surface mounting.

Junction boxes for ring final circuits

Spurs off ring final circuits may be connected by using junction boxes. The junction box usually employed for this purpose has three brass terminals in-line, each capable of taking up to four 2.5 mm², or three 4 mm² or two 6 mm² cables, or their equivalents. Junction boxes are normally provided with four side-knockouts for cable entry and two countersunk holes in the base for securing to joists, etc.

Where PVC-insulated twin-and-earth cables are used, the sheath of the spur cable should be stripped back for approximately 36 mm and the insulation removed for 12 mm. The ring final circuit cable can be prepared without actually cutting the conductors by carefully removing the sheath for approximately 62 mm, taking care not to cut into the core insulation, then removing a section of insulation approximately 12 mm long in the centre of the two unsheathed conductors. The two conductors and the bare earth wire are then laid into the terminal slots, the bared ends of the spur conductors laid on top of the appropriate main conductors and the screws firmly tightened. All non-sheathed portions of the cables must be within the junction-box enclosure. The cover is then screwed into place, any unused knockouts being blanked off with suitable insulating material. Spur connections may be made at right-angles to the box, or run parallel to the main cables.

Fitting 13 A socket-outlets

Flush-type socket-outlets are fitted into a box set into the wall or plaster in a similar manner to that already described for flush switches. Both metal and insulated moulded boxes are available, depending on the wiring system employed. Steel or cast-iron boxes to BS 1363 with fixing centres for single or twin socket-outlets should be used with metal-conduit systems. Knockouts for 20 mm conduit are provided. Metal boxes may also be used for sheathed wiring systems, in which case grommets or brass bushes must be fitted into the knockouts before cables are drawn in.

Where metal-conduit systems are employed, an earthing terminal must be fitted to the box and a bonding loop connected between it and the socket-outlet earth terminal (see Chapter 10, Safe and Efficient Earthing).

Figs. 5.5 to 5.7 show the steps involved in wiring a standard flush-pattern 13 A socket-outlet on a ring circuit.

Moulded pattress blocks can be obtained to enable flush fittings to be surface-mounted. These have both back- and side-entry knockouts for concealed or surface wiring.

Moulded surface-type socket-outlets are normally deeply recessed at the

Fig. 5.5. Wiring 13 A socket-outlets on a ring circuit. (1) After drawing into box each cable
should be stripped of about 32 mm of insulation

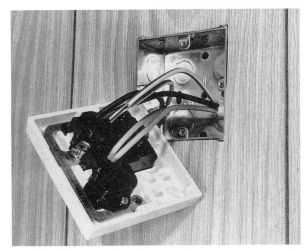

Fig. 5.6. (2) The tails are connected into the screw terminals. Connect the protective conductor last

back, and hence do not need a block if fitted to an incombustible surface. Where they are fitted to skirting boards, wood panelling or other combustible material, a pattress block or backplate must be used.

Metalclad, surface-mounted socket-outlets should be installed with surface metal conduit and in other positions where mechanical damage is possible. They are obtainable with brass or steel cover plates, the steel boxes normally being finished in aluminium stove enamel. Switched types are available with a pilot-light, if required.

For fixing and wiring a surface socket-outlet, the circuit cables are first passed through a suitable knockout in the pattress block, backplate or socket and the socket or block firmly screwed to the wall or other fixing position.

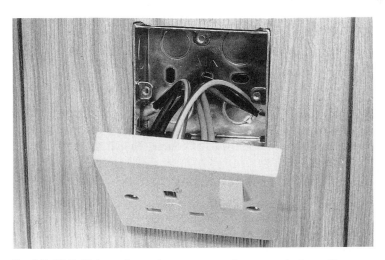

Fig. 5.7. (3) Hold the socket-outlet square on and secure to the box with screws

Slotted screw holes are normally provided and a spirit level should be used to ascertain that the fitting is level before final tightening of the screws.

If the cable is to be looped in and out, it should not be cut. The sheathing should be removed for approximately 75-100 mm, taking care not to cut into the core insulation, which should only be removed for a 32 mm length in the centre of the unsheathed portion. The bare conductors should then be doubled over for connection into the socket terminals. All unsheathed portions of cable must be within the pattress block or socket housing. If metalclad socket outlets are used, the metalwork of the box must be bonded to the earth terminal of the socket as previously mentioned.

The bare ends of the conductors are pushed through the appropriate terminals from the back and the clamping screws securely tightened. No bare cable should be visible at the back of the socket. With metalclad sockets, the switch and socket are attached to the cover plate or a separate grid for ease of connection.

When the cables and earth-loop wire (where applicable) have been connected, the socket is either screwed to the pattress or case with metal-thread screws into the tapped holes provided, or screwed direct to a wall or backplate with countersunk woodscrews. The slack cable should be neatly arranged in the pattress or case, so there is no tension on the terminals. When the socket is in position, the cover plate can be affixed with the metal screws provided (Fig. 5.8).

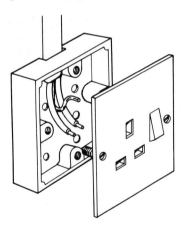

Fig. 5.8. A moulded pattress block can be used to convert a flush-fitting socket-outlet to a surface type

All socket-outlets in one room should be connected to the same phase. If the installation is fed by a single-phase supply only (which applies to most domestic dwellings), this condition will, of course, be met automatically.

Wiring 15 A, 5 A and 2 A socket-outlets

Although these sizes are now obsolete for new installations, they are still in fairly common use and are therefore included for reference when dealing with old wiring systems.

Socket-outlets of 15 A rating normally utilise a separate final circuit each, the wiring of each radial feed being capable of carrying at least 15 A. Socket-outlets have round pins and employ unfused plugs as each socket is fused at source. Switched types can be obtained with a pilot light. Both flush- and surface-mounted types are available, the surface type normally being fitted to a pattress block or backplate.

Socket-outlets of 5 A rating are assumed to be loaded at 5 A and, therefore, a maximum of three looped sockets are normally connected to each radial feed. The appropriate circuit protective device is 15 A and in this case when flexible cords are connected to 5 A socket-outlets the flexible must be of 15 A rating, unless fused plugs are fitted, with a smaller fuse. Single 5 A sockets on small radial feeds are often found to be connected with 5 A wiring and thus the circuit cannot be extended.

Socket-outlets of 2 A rating are normally assumed to be supplying a load of at least ½ A and are fitted for connection of light-current appliances, such as standard or table lamps. The circuit wiring is 5 A and any flexible cords should be capable of carrying the same current. 2 A socket-outlets are often found looped off local lighting circuits, but this practice should not be continued.

Electric clock outlets

Electric clocks *can* be connected via socket-outlets to BS 1363 and fused plugs rated at not more than 3 A. It is preferable, however, to use special fused connector boxes, which are available in surface-or flush-types and have a flex-outlet on the front. They are fitted with 2 A or 3 A cartridge fuses which can be replaced by removing the centre section of the front cover. The flush-type can be fitted to plaster depth metal boxes. Many electric clocks have double-insulated wiring and hence do not require an earth connection, but the fused connector boxes can be wired with earthing facilities if required.

The wiring may be run as a separate radial feed from the distribution board, or looped off a convenient final circuit at a fitting or via a junction box. In all cases, a fuse must be incorporated at the socket-outlet or connector box.

Switch control of socket-outlets

Switched-sockets are often convenient to the user and should normally be fitted in preference to unswitched. Socket-outlets on d.c. supplies *must* be controlled by an adjacent switch, but with a.c. supplies the switch is optional.

The standard 13 A switched socket-outlet normally has a single-pole switch, but 13 A and non-standard sizes, such as 5 A and 15 A, can be obtained with double-pole switches. The single-pole switch must always be connected to the phase terminal of the socket-outlet, so that, provided the socket is correctly wired, the phase conductor is broken.

Socket-outlet adaptors

The use of adaptors should be avoided as far as possible, and this can be achieved by providing an adequate number of socket-outlets, as indicated earlier. If adaptors are used with 13 A socket-outlets to BS 1363, the outgoing circuits need not be fused in the adaptor since fuses will be provided in the plugs on the appliances. However, a fuse not exceeding 13 A should be provided in the main incoming phase connection of the adaptor since otherwise the adaptor and the 13 A socket feeding it could be seriously overloaded, e.g. two 13 A loads could be plugged into the adaptor.

On adaptors used with other types of socket-outlet, where the plugs are unfused, any outgoing circuits of lower rating than that of the input should be fused.

Flexible cords for portable apparatus

Flexible cords for use with portable appliances may have many different types of insulation, depending on the application and operating conditions. Most flexible cords on domestic apparatus have vulcanised- or butyl-rubber or PVC-insulation.

The colour identification of the cores of flexible cords and flexible cables is: brown for the phase conductor, blue for the neutral and green and yellow for the protective conductor.

The current rating of a flexible cord must be not less than that of the fuse or circuit-breaker through which it is fed.

Flexible cables used in high ambient temperature conditions, e.g. for the connection of wash-boilers, immersion heaters or radiators, should have heat-resisting insulation such as butyl or silicone-rubber.

Provision for television and telephones

In some cases it may be desirable for the electrical installation to include provision for the wiring of television aerials and telephones to avoid unsightly surface wiring at a later date. It is most likely to be required in blocks of flats but it can also be useful in normal housing, particularly in the more expensive type.

For television, the provision usually consists of a concealed conduit, with draw wire, from each aerial outlet position to a suitable point where the aerial can be picked up. This point will usually be the roof space, in a house, or the vertical riser duct or other suitable point in a block of flats.

Where a communal television aerial serves a number of flats, a 13 A socket may also be required at a suitable point to feed the common amplifying equipment. Specialist advice on the requirements should be obtained when necessary.

At the aerial outlet positions the conduit should terminate on flush type outlets designed for the purpose; these are readily available. A 13 A socket-outlet should, of course, be adjacent.

Similar provision may be required for telephones, and details should be agreed with the appropriate authority.

The above conduits should, of course, be quite separate from any other conduits.

6 Wiring fixed appliances

Depending on the electrical loading, fixed appliances may be connected either (a) to ring or radial circuits feeding 13 A sockets to BS 1363, or (b) on individual circuits from the consumer unit.

The former category includes, fixed fires, fans, *small* water heaters and hand driers, and these should be connected via fused connection units — see Chapter 5.

The second category covers floor-standing or other cookers rated at more than 3 kW, water heaters, off-peak storage heaters and other large loads. Each item should be on a separate circuit and be controlled by an adjacent switch; this need not normally be fused since overload protection will be provided at the consumer unit.

Two stationary cooking appliances installed in the same room in domestic premises may be controlled by the same switch provided that neither appliance is more than 2 m from the switch. Every stationary cooking appliance must be controlled by a switch separate from the appliance.

Cooker controls

The rating of a cooker circuit should be calculated in accordance with the details in Chapter 2, Table 2.2. A 30 A or 45 A circuit is normally necessary and the wiring should terminate on either a cooker control unit (which includes a socket-outlet) or a simple 30 or 45 A switch. Control units and switches are available with pilot lamps if required. Fig. 6.1 shows typical units and Fig. 6.2 shows the internal wiring of a cooker control unit. Note that the socket-outlet is not controlled by the cooker switch. The 13 A socket-outlet is not fused within the control unit since plugs used with the socket will include fuses. It is preferable, and more usual these days, to use a simple switch rather than a cooker control unit since the inclusion of a socket-outlet can cause danger due to flexible cables trailing over the top of the cooker.

The switch or control unit should preferably be mounted at the side of the

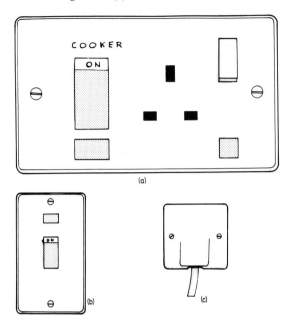

Fig. 6.1. (a) Cooker control unit. (b) 45 A switch. (c) Cable outlet (*Tenby Electrical Accessories Limited*)

cooker, rather than behind it, at approximately 1350 mm (minimum) above floor level.

Surface-mounted cooker-control units have between five and seven knockouts in the sides and back for incoming and outgoing cables. Flush units are fitted into a steel box which is sunk into the wall. The box has 25 mm knockouts in the sides and back for incoming and outgoing cables. These

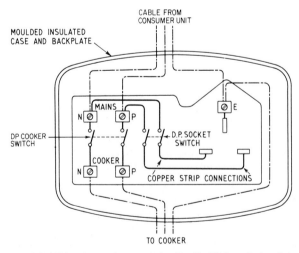

Fig. 6.2. Wiring for cooker control unit with 13 A socket-outlet

should be fitted with a rubber grommet or a bush when sheathed wiring systems are employed.

The interiors of cooker control units or switches are either mounted on the faceplate or on a separate removable chassis. When a surface box is used the cable to the cooker emerges from the bottom of the unit and passes direct to the entry point on the cooker.

If the connection unit is a flush type it is usual to run a short length of concealed conduit from the unit to a point lower down the wall opposite the entry point on the cooker, and at this point a special flush cable outlet (Fig. 6.1(c)) is fitted. The cable is then run via the conduit from the control unit to the cooker. The final connection to the cooker need not be flexible cable and if the cooker circuit is wired in PVC-sheathed flat twin-and-earth cable, the same cable can be used from the control unit to the cooker.

The free cable between the control unit and the cooker should be long enough to allow the cooker to be moved for cleaning and maintenance. It is preferable that a cooker should not be sited adjacent to a refrigerator, and the control unit should preferably be at least 2 m from a sink, particularly if it includes a socket-outlet.

Three-heat switch control

Some electric cookers are provided with a three-heat switch (Fig. 6.3). When the switch is in the 'high' position, the two heating elements, usually rated at 1,000 W each, are connected in parallel across the supply. When the switch is turned to the 'medium' position, only one element is energised, so that the oven or grill is now operating at half its full rating. When the switch is turned to the 'low' position, the two elements are connected in series across the supply; a simple calculation will show that the power is again reduced by one half. Thus, the heating effects produced by the three-heat switch are in the following proportions:

high — full rating, say 2,000 W
medium — half rating, say 1,000 W
low — quarter rating, say 500 W

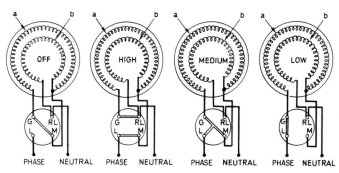

Fig. 6.3. Connections for three-heat switch. Off—elements a and b are disconnected; High—a and b are in parallel; Medium—a is short circuited but in series with b; Low—a and b are in series

'Simmerstat' control

Another method of heat control for boiling plates is the energy regulator, or 'Simmerstat' controller. This is designed to provide infinite regulation of the input to boiling plates.

The principle is the breaking of the circuit at variable intervals of time by means of a bi-metallic strip (see Fig. 6.4). If a boiling plate is switched on for,

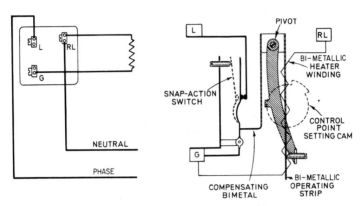

Fig. 6.4. Construction and connections for type TYC 'Simmerstat' control

say, 10 seconds and then switched off for 10 seconds repeatedly, over a period of time the amount of energy consumed is half that consumed if it is switched on all the time, giving half or 'medium' heat (see Fig. 6.5). If, the ratio of on and off periods is altered to give five seconds on and fifteen seconds off, the result is quarter or 'low' heat.

The regular on–off switching cycle is obtained by the bending of a bi-metallic strip under the influence of a smaller heater, which is wound on the operating bi-metallic and is arranged for shunt connection across the load. The percentage time 'on' during the switching cycle is governed by the relative position of the bi-metallic assembly, which is altered by the movement of a cam on the control-knob spindle. The knob is free to rotate 360° in either direction, so any desired setting can easily be obtained.

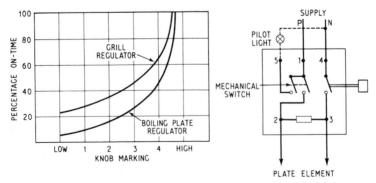

Fig. 6.5. Performance curves for 'Simmerstat' type controller and (right) connection for controller with mechanical switch and pilot light (Satchwell)

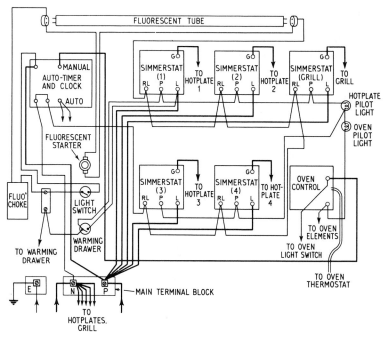

Fig. 6.6. Cooker control layout, including fluorescent tube and warming drawer

Certain types of Simmerstat controllers have a separate mechanically operated switch in addition to the bi-metal, both breaking the circuit when the knob is in the off position, so that boiling plates are isolated from both supply lines. A separate pilot-light switch is often incorporated, arranged so that several controllers can be connected to the one pilot-light (Fig. 6.6).

When the switch is set to 'high' the heating element is energised continuously, while at intermediate knob settings the power is automatically reduced by the periodic switching. At the minimum or 'low' setting, the input is approximately 5 per cent of the rated loading of the heating unit. A slightly different controller is usually fitted for grills which gives an increased input at low-knob positions.

Oven thermostats

There are two types of oven thermostat used by cooker manufacturers, the capillary-tube type and the bi-metallic type.

Capillary-tube thermostat This is the most widely-used type of oven thermostat, slightly different versions also being used for washing machines and boilers.

The thermostat consists of a thermally-sensitive phial, approximately 5 mm in diameter, connected to the switch unit by a length of metal capillary tubing (see Fig. 6.7). Changes of temperature of the liquid in the phial are transmitted hydraulically through the capillary tube to a capsule in the switch

Fig. 6.7. Satchwell TO capillary tube thermostat for oven control

unit. Movement of this capsule actuates a micro-gap switch when the desired temperature set by the control knob has been reached. An additional switch is sometimes fitted to provide double-pole isolation in the off position. The liquid in the phial is sensitive to temperature changes between 100 and 300°C. Ambient-temperature compensation is incorporated to prevent changes in calibration due to heating of the switch unit. An adjustment screw is usually provided on the switch unit to apply a 'false' setting up to 28°C to the control knob.

Some ovens are provided with separate 'bake' and 'grill' elements, and a switch head can be obtained which incorporates a separate switch for each.

Bi-metallic tube thermostat The bi-metallic tube is almost identical with the water-heater thermostat and consists of a brass tube which forms the heat-

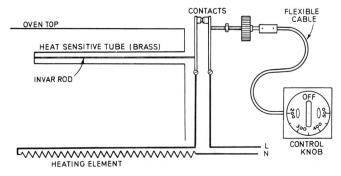

Fig. 6.8. Principle of bi-metallic oven thermostat

sensitive part of the instrument, the free end being fastened to a rod inside the tube (see Fig. 6.8). The rod is made of a metal alloy which is hardly affected by heat, so that expansion or contraction of the brass tube due to heating or cooling will be fully transmitted to the end of the rod, where it operates contacts through a lever. The contacts work on the micro-gap principle, so only a small movement of the brass tube is required to open or close them.

Various devices are used to make the action positive, such as by fitting a small permanent magnet close to the contact arm, and there are different ways of varying the setting of the thermostat. The ultimate effect is the same, i.e. the contacts make or break at a predetermined movement of the heat-sensitive parts.

Fires and heaters

Heaters for unrestricted local heating (i.e. not off-peak) can be divided into three categories: radiant, convector and fan.

Radiant heaters

Usually, these have a wound element on a ceramic rod, which glows at red heat. The heat is reflected into the room by a parabolic polished-metal reflector. Elements are normally 0.75 or 1 kW and up to four 0.75 kW or three 1 kW elements can be fitted to each fire. Radiant heaters can also be obtained with wound elements enclosed in silica glass; this type is often used in kitchens, bathrooms or other damp situations.

Radiant heaters can be free-standing, wall-mounted or inset (built into the wall). Free-standing heaters are normally connected, via a flexible lead and plug, to a socket-outlet. Wall-mounted and inset heaters should be connected through a fused connection unit as described later.

Convector heaters

These can be divided into three groups: conventional convectors, oil-filled radiators and tubular heaters.

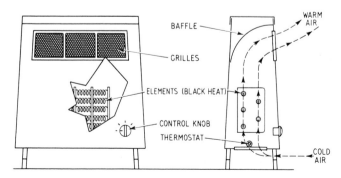

Fig. 6.9. Construction of typical convector heater

Conventional convectors consist of a styled metal case with a cold-air inlet at the bottom and a hot-air grille at the top. The heating element operates at black heat, thus no reflectors are necesary and the casing of the heater does not become very hot. The heat output can be varied manually, or the heater can be fitted with a thermostat which will switch the heater off when the air entering the bottom of the heater is at the required temperature (see Fig 6.9).

Oil-filled radiators have an enclosed element running along the bottom to heat the oil; they may be manually or thermostatically controlled, and loadings of 0.5 to 2 kW are available. They can be fixed to the wall or free standing.

Tubular heaters comprise a black-heat element inserted in a steel tube of approximately 50 mm diameter. Tubes can be obtained in lengths of 600 mm to 5 m and can also be banked where space does not permit the required loading in one tube; they may be fixed or free standing (see Fig. 6.10).

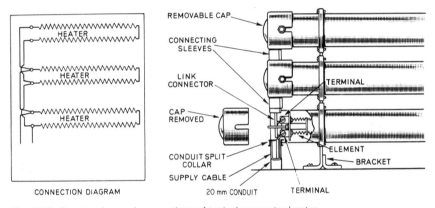

Fig. 6.10. Construction and connections of typical convector heater

The heater elements of most convector-type heaters are wound with nickel-chrome alloy wire. The tubular heater usually has a rating of 130-300 W per metre length, the rating being selected which best suits the installation. For instance, when a room has a glass roof or skylight, it is possible to stop the down draught by tubular heaters fitted round the glass. The loading per metre is usually increased to about 300 W in this case.

Provided suitable heat-resisting paint is available, any colour may be used for decorating radiator- or tubular-type heaters. For highly-polished nickel- or chromium-plated tubular heaters, the loading must be reduced to about 150 W per metre if the normal surface temperature is not to be exceeded.

Fan heaters

Fan heaters consist of a metal case enclosing black-heat elements and a motor-driven fan. The fan may be of the centrifugal or tangential type and is often two-speed which, in conjunction with variable switching of the heater elements, gives a controlled output. Heaters may be free-standing or wall-mounted, but for maximum effect should operate at floor level.

Thermostats for convector heaters

Most convector-type heaters may be thermostatically controlled, either by integral or separate wall-mounted thermostats (see Fig. 6.11). The purpose of a thermostat is to maintain an equable temperature within a room, but as the temperature often varies for different parts of the room, wall-mounted types should be fitted to deal with average values. While thermostats should be freely exposed to the air, they should not be fitted in positions subject to draughts or radiation from any heat source.

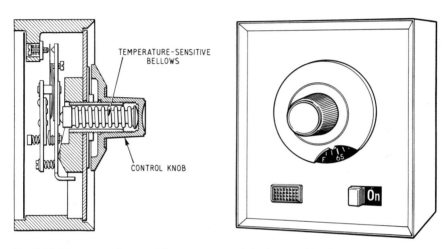

Fig. 6.11. Thermostat for use with convector or tubular heaters. The thermostat can handle loads up to 20 A and is connected between the supply and the heater elements (*Maclaren Controls Limited*)

If the space to be heated is large, it is usual to split up the total heating load in each room into sections of up to 3 kW, a separate thermostat being fitted for each section.

Many thermostats employ an accelerator heater which causes the thermostat to anticipate a rise in temperature when it is 'calling for heat', thus reducing temperature overshoot.

If the thermostat does not have a definite 'off' position, a separate or integral switch must be incorporated with the heater.

Preset-time control of heaters may be used as an alternative to, or in conjunction with, a thermostat. For this purpose an electric synchronous motor-driven clock with two or more 'on' and 'off' switching cycles is used.

Such clocks are available with spring reserve so that they continue to operate for up to 24 hours if the mains supply is interrupted.

Size and position of heater

When deciding on the size of radiant or convector heater or heaters that will be required for a certain-size room, it is usual to work on the asumption of 35 W/m³ of room space. This figure is obviously approximate, since a room with a large amount of window space will require more heat input than a room of the same dimensions having small windows. Similarly, a room having two or three outside walls will require correspondingly more heat than one with only one outside wall. It is, however, seldom necessary to exceed 50 W/m³.

Where convector heaters are used to heat a room, between 35 and 45 W/m³ of air space is required, depending on local conditions. If, however, the convector is thermostatically controlled, a slightly better arrangement is to use a heater with a loading well in excess of the calculated minimum. A room that is continually being used and is heated by a convector is usually more economically heated if a thermostat, possibly with preset-time control, is used.

Radiant heaters are normally installed in positions where direct heat is required at the focal point in a room, e.g. in a fireplace to replace a coal fire, or high up on a bathroom wall, angled to direct the heat towards a person standing beside the bath.

Convector heaters of all types are more often found in cold corners of a room, or under windows to prevent cold draughts. Where heaters are installed under windows, they should not be screened behind long curtains. Apart from the risk of fire, the effect will be to drive most of the heat through the window and isolate the heater from the room.

With all electric heaters, the temperature limit is reached only when the heat loss is equal to that generated. If, therefore, heaters are in any way blanketed by shelves, furniture or fittings, the surface temperature will be higher than for freely-exposed units.

Wiring a fire or heater

Portable heating appliances connected by a flexible cable and plug to a socket-outlet are normally of metal construction, and therefore three-core flexible cables and three-pin socket-outlets must be used. Care should be taken to see that the protective conductor is properly connected.

Fixed heaters, either wall-mounted or inset, must be connected through a fused connection unit to a ring or radial final circuit if they are permanently wired (i.e. not via a plug and socket) — see Chapter 5. However a heater which is on a separate circuit from the consumer unit may be controlled by a simple unfused switch.

Each heating appliance must be controlled by a switch which disconnects the phase conductors. This switch may be separate from the heater, e.g. on the connection unit, provided this is readily accessible, or it may be on the heater itself provided the connections are arranged so that the appliance can be dismantled for maintenance without exposing any parts which remain live when the switch is open.

Heating appliances where the heating element can be touched must have a *double-pole* switch which disconnects *all* the circuit conductors. IEE Regulation 601-12-01 requires that the sheath of a silica-glass sheathed element should be regarded as part of the element and the above requirement therefore applies.

All metallic non-current-carrying parts of heating appliances must be electrically connected together and to the protective conductor of the final circuit; an earth terminal is usually provided for this purpose. Where conduit systems are employed, a bonding wire should be connected between the conduit and the earth terminal of the heater.

Table 6.1. Loading for radiant and convector heaters

Situation	Loading in kW per 30m³		
	Rooms with one exposed wall	Rooms with two exposed walls	Rooms with three exposed walls or with an excessive area of glass
Houses			
Ground floor with no basement	1–1¼	1½–1¾	1¾–2
Intermediate floor	1–1¼	1¼–1½	1½–1¾
Top floor	1¼–1½	1½–1¾	1¾–2
Bungalows	1½–1¾	1¾–2	2–2¼
Intermediate passages	Allow 1 kW per 30m³		

Note: With an outside temperature of 0°C these loadings will give an internal temperature of 18°C. If an internal temperature of 24°C is required, increase the loadings by one-third. If an internal temperature of only 13°C is required, these loadings can be reduced by one-third. (Dimplex Ltd.)

Fixed heaters are usually provided with a metal barrier to prevent fixed wiring from becoming too hot, the connections on the heater itself being made with asbestos-covered or porcelain-beaded wire. If the fixed wiring temperature is likely to exceed that specified in the IEE Regulations Appendix 4, sleeves of heat-resisting insulation should be used, e.g. silicone-bonded glass braiding.

Fitting inset fires

If an inset fire is screwed to a wall without a surround of some sort — e.g. tiles, marble, slate or asbestos-cement sheet — discoloration of the wall surface above the heater will take place. The actual methods of securing the heater to the surround vary a great deal. One method is to cement the heater box in the opening and fix the front to the box by screws after the electrical connection has been completed. Another method is to fit two spring arms on the back of the heater which are so spaced that they grip the sides of the opening in the surround into which the heater is fitted. Quite a number of the cheaper fires are secured by plugging the surround and screwing through the front. A good method is shown in Fig. 6.12.

When electric fires are fitted in fireplaces, see that there is not an open flue immediately above the fire, otherwise heat emitted by convection from the back will pass directly into the flue and be lost. While some form of ventilation other than windows is very necessary in all rooms, the average chimney is more than sufficient. The flue should be partly blocked and the air entry lowered to below the level of the top of the fire.

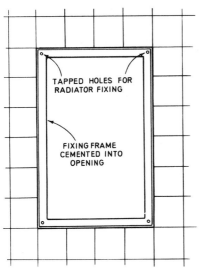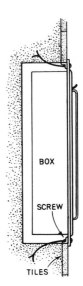

TAPPED HOLES FOR
RADIATOR FIXING

FIXING FRAME
CEMENTED INTO
OPENING

BOX

SCREW

TILES

Fig. 6.12. Typical method of fixing an inset fire

Special precautions for heaters in bathrooms

IEE Regulation 601-12-01 requires that a heating appliance having heating elements which can be touched is not to be installed within reach of a person using a bath or shower. As stated earlier, the sheath of a silica glass sheathed element is regarded as part of the element. Any fixed heater must have a cord-operated switch or a switch outside the room. Portable heating apparatus should not be taken into bathrooms and no socket-outlets must be installed, except special shaver sockets, as described in Chapter 5.

Heaters recommended for use in bathrooms are the wall-mounted, cord-operated 'infra-red' radiant type. These have a silica-enclosed spiral element and the angle of the reflector is adjustable. Once the heater has been screwed to the wall and connected, the reflector can usually be adjusted to the required angle and then locked in position by tightening a small screw.

Oil- or water-filled, electrically heated towel rails in bathrooms should be connected by means of a fused connection unit. As they do not normally incorporate a switch, a separate means of control must be fitted. This can be effected into two ways:

1   By inserting a junction box in a suitable power circuit, or looping from an outlet position, and running a twin-and-earth cable (or two singles if a conduit system is employed) to a switched, fused connection unit (preferably with a pilot light) mounted just outside the bathroom door. From the connection unit another cable is run to a convenient position adjacent to the towel rail and terminated at an unfused flex-outlet plate from which the heat-resisting flexible cable is taken to the towel rail (see Fig. 6.13).

2   By using an unswitched, fused connection unit fitted in a convenient position outside the bathroom and running a cable from this to a cord-

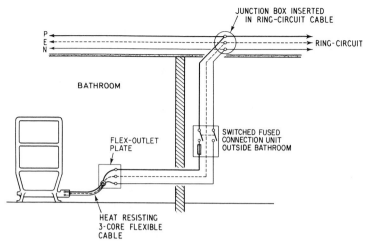

Fig. 6.13. Connecting an electrically heated towel rail using a switched fused connection unit and an unfused flex outlet plate

operated ceiling switch inside the bathroom. The switch should be 15 A double pole, preferably with a pilot light. From the switch a cable is run to a flex-outlet plate as before.

IEE Regulations 601-04 requires that in a room containing a bath or fixed shower, exposed conductive parts must be bonded together and earthed. This covers the metalwork of such items as heaters, towel rails, radiators, metal baths or sinks, pipework etc. The bonding should be by the correct size of single-core cable and suitable bonding clamps or clips — see Chapter 10.

Off-peak systems for central heating

Electric central heating can be carried out by three basic methods:

1 Underfloor heating
2 Storage radiators
3 Warm-air circulation (Electricaire).

All three systems use the thermal storage principle, whereby a large mass of heat-retaining material is heated during the electricity board's off-peak periods and allowed to emit the stored heat throughout the day.

The Generating Company's generating plant has to be of sufficient size to cater for the maximum day-time demands of industrial and domestic consumers. Much of the plant must be running for 24 hours a day and at certain times there is a good deal of spare capacity in the system. At such off-peak times, the electricity boards sell the electricity at reduced price to encourage consumers to make use of this spare capacity.

Heating systems have therefore been developed which store heat during off-peak periods and emit heat during the day. With a separately-metered off-peak tariff the charging period is normally at night, but with some systems a midday boost is required to ensure an even heat emission until evening. However a day-time boost is not now normally available to new domestic consumers. This off-peak tariff is known as 'Economy-7'.

Floor-warming installations

Floor-warming utilises the thermal storage properties of concrete — the concrete floor of the building being heated by special cables. These cables are embedded in the concrete screed just below the floor surface and when a current is passed through them, the high resistance of the conductor causes it to become heated. This heat is conducted into the concrete which in turn radiates the heat into the room. The system lends itself to operation on the off-peak supply since the concrete, once heated, will continue to give off heat for a considerable time after the supply has been switched off. The current input is usually controlled by one or more thermostats.

The fact that the cable must be solidly embedded in the concrete floor screed means that its use is virtually restricted to new buildings or major reconstruction of old ones. For this reason also, it can only be installed on the ground floor in conventional domestic premises (with the notable exception of flats and maisonettes).

Installing the heating cables

Where floor-warming cables are to be installed, the floors of a building must be put down in layers. First, the hardcore or aggregate is covered by a layer of concrete and allowed to harden. This is known as the sub-floor. A damp-proof membrane and perimeter insulation should also be applied to the ground floor slab to prevent undue heat loss; a typical cross-section is given in Fig. 6.14. Expanded polystyrene or a similar material can be used to provide thermal insulation at the floor slab perimeter.

In each area to be warmed, cable spacers are then fixed to the sub-floor concrete 150 mm from, and parallel to, two opposite walls, usually the shortest sides. These spacers may consist of saucer-shaped metal discs which are attached to the floor with a Hilti tool (a device which 'fires' a metal pin into the concrete), or special plastic spacer bars. A typical plastic spacer bar is shown in Fig. 6.14. This type is manufactured in 1800 mm lengths and is permanently fixed to the sub-floor by plugging the floor and using small woodscrews. For large areas, additional spacers should be fixed at a maximum of 6 m intervals to ensure that correct spacing of cables is maintained during screeding. The length of cable run should not in any case exceed 9 m without a change in direction.

Floor-warming cables are supplied in fixed-loading units from ¼ to 5 kW, their length depending on the type of insulation and kilowatt loading (see Fig. 6.15). Suitable combinations of units can be used to suit any particular requirement. Loading varies with insulation and sheathing, but values of

10-16 W/m length are common. Many heating cables are colour-coded to indicate their kilowatt loading and the length of each unit supplied is indicated.

By measuring the dimensions of the area to be heated, the spacing of the cable runs can be calculated to give an even distribution over the whole floor.

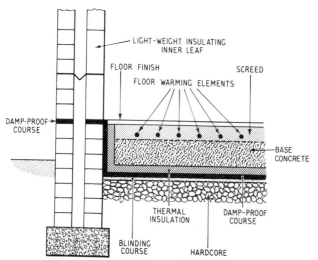

Fig. 6.14. Installing a floor warming cable. The cable is laid onto concrete and plastic spacer bars (above) are used to make the turns at the end of each run (*BICC*)

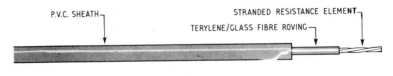

P.V.C. SHEATHED FLOOR WARMING CABLE

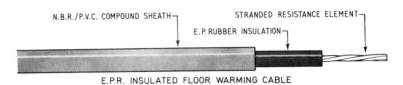

E.P.R. INSULATED FLOOR WARMING CABLE

Fig. 6.15. Typical floor warming cables

In the average installation, the cable runs are spaced at a distance of 75-150 mm.

The cable is laid direct on to the sub-floor from spacer to spacer across the room with a gentle U-turn round the spacers. It should not be hammered or forced into position and care should be taken not to tread heavily on it or drop heavy objects on to it. A monitoring device or a high-resistance testmeter such as a 'Megger' should be connected across the cable cores during installation to ensure that any damage is detected before final screeding is carried out.

Connecting the heating cables

Floor-warming installations can be considered as large fixed appliances and a switch must be provided to control them.

The heating cables are normally supplied with 'cold tails'. These are lengths of ordinary 1.5 or 2.5 mm² sheathed wiring cable which are fixed to the ends of the heating cables at the factory. Where heating cables are supplied without cold tails, they must be terminated in a special box which can be buried in the wall at skirting level. The cold tails are then taken to a thermostat which can be mounted in a suitable position in the room. The thermostat should preferably be the type that has definite on/off positions.

Fig 6.16 shows how the cables from the thermostat may be connected either to a fused connection unit or run back to the main distribution board or consumer unit. If miniature circuit-breaker distribution boards are used, then a separate on/off switch in each circuit is not required.

Floor-warming circuits are usually controlled by a separate main switch and connected to a separate distribution board or consumer unit with fuse or MCB overload protection. The system is normally connected on the off-peak supply and a contactor is usually necessary since the load is normally too large for a timeswitch only.

Each separate floor-warming cable is usually controlled by its own thermostat, although small rooms where background heating only is required may be heated by one continuous cable and controlled by one thermostat.

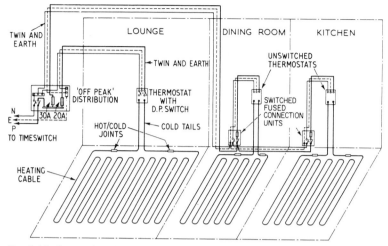

Fig. 6.16. Connection of floor warming cables showing use of switched thermostats and switched fused connection units

The setting of the thermostat determines the length of time that each cable is supplied during the available charging period. The thermostat may incorporate an on/off double-pole switch, or a separate switch or MCB may be used to isolate each cable when not required. Each separate floor-warming cable must be connected to a separate way of the off-peak consumer unit, or to a local fused connection unit as shown in Fig. 6.16. IEE Regulation 601-12-02 requires that, for a bathroom, the floor heating should be covered with an earthed metallic grid and connected to the local supplementary bonding.

Off-peak storage radiators

The off-peak storage radiator is a free-standing metal cabinet which contains a number of refractory blocks, heated by elements which are inserted between them or embedded in them. The blocks are heated during the off-peak periods, and are encased in insulation which allows a continuous constant heat emission throughout the whole day.

All storage heaters have a controlling device which adjusts the input charge, this being effected by the manual adjustment of a thermostat to suit the prevailing outside temperature.

On some heaters there is no control of the heat output; on others the output can be controlled, manually or automatically, by a damper. Alternatively the output is controlled by a fan which is in turn controlled by a switch on the heater or a room thermostat and, in some cases, a timeswitch. The fan-assisted type (Fig. 6.17) usually has extra insulation to restrict the heat output when the fan is not in use. Modern off-peak tariffs usually make the supply available for a 7-hour period during the night only and a daytime boost is not normally available. Modern storage heaters are designed to cater for this condition with a capacity to cover the full output period required. Loadings

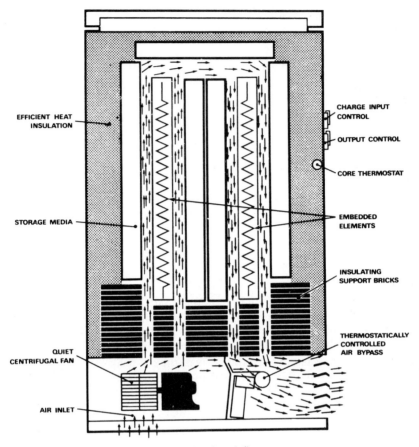

EFFICIENT HEAT
INSULATION

CHARGE INPUT
CONTROL

OUTPUT CONTROL

CORE THERMOSTAT

STORAGE MEDIA

EMBEDDED
ELEMENTS

INSULATING
SUPPORT BRICKS

THERMOSTATICALLY
CONTROLLED
AIR BYPASS

QUIET
CENTRIFUGAL FAN

AIR INLET

Fig. 6.17. Fan-assisted storage radiator showing air flow

of up to 3.4 kW are now available and it should be noted that 13 A fusing is not satisfactory for this load.

The rating of a storage heater is of course the input loading, in kilowatts, of the heating elements.

Radiators are available from 1¼ kW to 3.4 kW and the number and sizes of radiators required can be roughly calculated from Table 6.2. The number of watts per square metre must be multiplied by the floor area of the room to give the total number of watts required. Dividing by 1000 gives kilowatts.

Example: A living-room in a normal house, measuring 5 m × 4 m, required temperature being 18°C

$$\text{Load} = 5 \times 4 \times 160 \text{ W}$$

$$= 3200 \text{ W}$$

$$= 3.2 \text{ kW}$$

Therefore to heat this room to 18°C one 3.4 kW storage heater will be required.

Table 6.2. Size of storage heater required for different rooms

Type of room to be heated	Required temperature (°C)	Watts per square metre
Normal rooms with not more than two exposed walls and ceiling height of 2.13–2.74m	18	160
	13	110
Rooms with ceiling height above 2.74m, or with three exposed walls, or modern rooms with large windows	18	210
	13	160
Single rooms where not a part of general heating	18	210
	13	160

This method is only approximate and, in case of doubt, advice should be sought from a manufacturer, or the electricity board.

Positioning and installing storage radiators

The most effective results will be obtained if a storage radiator is installed to radiate heat across a window rather than being sited directly below one. The heater should not be installed in such a position that large horizontal temperature gradients across the room will be experienced, e.g. against an interior wall in a room having a window facing north.

When the position and size of each heater has been determined, the wiring circuits can then be planned. When determining cable sizes and voltage drop, a 3.4 kW load at each outlet point should be allowed for, irrespective of the size of heater being installed. This will allow for future changing of heaters.

Fig 6.18 shows different arrangements which may be adopted when wiring for storage radiators. A single outlet or two outlets may be wired off each

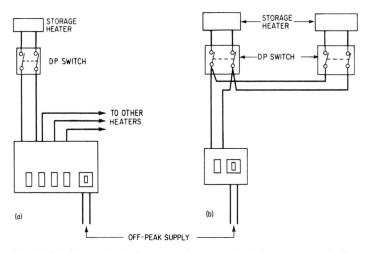

Fig. 6.18. Wiring methods for storage heaters, using (a) consumer unit, (b) splitter unit or a switchfuse as required

circuit, using suitable ratings for cables and switchgear. No diversity can be allowed since all heaters are switched on and off together.

Single heaters wired to a separate way of the consumer unit may be controlled at the outlet point by a 20 A unfused switch, preferably the type with a flex outlet on the switch plate; the distribution fuses or circuit-breakers should be of 20 A rating.

Where two heaters are wired on one circuit, the fuse or circuit-breaker in the consumer unit, and the wiring, should be of 30 A rating. A double-pole switch, preferably with flex outlet, should be provided adjacent to each heater.

Fan-assisted storage radiators

Where fan-assisted storage radiators are installed, the heater circuit is wired through the off-peak consumer unit as described above, but the fan must be connected to a normal unrestricted circuit to ensure that it can be switched on at any time.

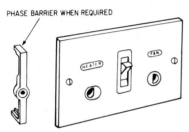

PHASE BARRIER WHEN REQUIRED

Fig. 6.19. Switch for storage radiator supplies using 25 A d.p. switch controlling both feeds (*MK Electric Limited*)

This involves two separate feeds to the radiator, and provision must be made for these to be isolated when required. Special switch units are available for this situation and they include a d.p. switch which cuts off both supplies and also two flex outlets for the flexible cables to the radiator. Fig. 6.19 shows a typical unit which includes a 25 A d.p. switch. If the two feeds are on different phases, a phase barrier and suitable warning labels must be added to the switch; however this situation is not very common on domestic installations since the supply is normally single phase.

An alternative type of switch is a combined unit with separate switches for heater and fan. The heater switch is 30 A and the fan 5 A. A 5 A fuse may also be incorporated with the fan switch to permit the fan to be connected to any convenient power circuit. If a fused type of fan switch is not used, the fan supply must be brought back to a 5 A way on the unrestricted supply consumer unit, or fed from a power circuit via a fused connection unit. However, as stated in IEE Regulation 537-02-07, a common switch cutting off both supplies is preferable.

The fan on some fan-assisted storage radiators may be remotely controlled by a thermostat and/or a timeswitch. In this case, looping terminals are provided on the switch unit for connection to the thermostat or timeswitch or both.

Twin-and-earth PVC-sheathed cable should be used throughout for the fixed wiring to the fans and heaters, unless a conduit system is used. The final connection to the heater should be with a 3-core butyl or other heat-resisting flexible cable.

Assembly of storage radiators

Storage radiators are normally delivered with the central core of refractory blocks in separate containers. Assembly is usually effected by removal of the front and top, or back and top, which are held by small bolts or self-tapping screws. The flexible heat-resisting cable is then passed through the bushed hole provided and connected to the mains terminal block. Provision is usually made for the cable to enter at either end.

If the heater is fan-assisted, the flexible heat-resisting lead for the fan must also be brought into the bushed hole provided and connected to the switch terminals. Where the fan is to be automatically controlled, two more terminals are provided on the fan switch and a connection to the thermostat or timeswitch must be made at this stage.

The heater case must then be placed in position, as it is very difficult to move storage radiators once the heating blocks are in place. If the insulating material is contained within the heater case, the heating core will probably comprise several small refractory bricks, each weighing 7 to 9 kg. These have the heating elements embedded in them and have two heat-resisting insulated wires for connection. Some types of storage radiators have a sheet-metal case and heating blocks contained in two or more metal canisters, which are simply lifted into position on the heater base. The canisters are again provided with two insulated, heat-resisting connecting leads. The connecting wires for the blocks or canisters are connected either to busbars within the heater or interconnected porcelain-insulated terminal blocks.

The internal wiring of the heater, connecting the input control unit and the automatic cut-out and internal thermostat, is normally carried out by the manufacturer.

When the case has been reassembled, the flexible lead or leads can be connected to the flex-outlet switch. The cord-grip device provided should be firmly tightened before the switch is screwed into position.

Electricaire ducted warm-air system

Electricaire is the name given to the Electricity Council's Specification for ducted warm-air central-heating systems; many manufacturers make heaters which conform to this specification.

The Electricaire system is based on a centrally sited thermal-storage heater with a high storage capacity and very good insulation to retain the stored heat until it is required (see Fig. 6.20). As with all electric central-heating storage systems, the unit is 'charged' during the electricity board's off-peak period. A built-in fan, which can be controlled either by hand or automatically by room thermostat or timeswitch, delivers the stored heat in the form of warm air. The warm air is mixed with cool air as necessary to maintain the required temperature and is then carried from the central unit through concealed ducts

Fig. 6.20. Creda Electricaire heating unit

to outlet grilles in the walls or floors in any part of the house. This system is therefore normally installed in new properties at the construction stage, although it can be installed in existing properties provided there is a sufficient cavity beneath the floorboards to conceal the ducting.

The size and capacity of the Electricaire unit varies according to the type and output required, but in most cases it can be accommodated in a cupboard approximately 750 mm square. For most domestic properties a 6-8 kW unit is sufficient and for larger properties, units up to 15 kW are available.

The outlets in the rooms may have variable dampers which can be manually adjusted to control the output of warm air. The fan speed can also be increased by a special manual switch to boost the output.

Construction and installation

The first sections of the Electricaire unit must be installed when the dwelling is at the damp-course stage of construction. Once the position of the heater has been established the chamber on which the heater will rest, and to which the outgoing ducts are attached, can be installed. This is known as the plenum chamber and is normally constructed with an angle-iron frame and 22-gauge sheet-metal cladding. The plenum chamber is usually laid on a small concrete pad so that the top will be at final floor level. The flanges for attachment of

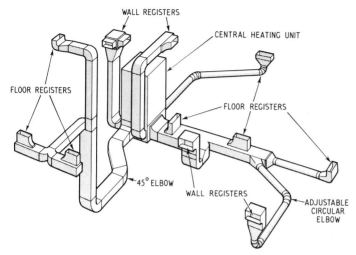

WALL REGISTERS

CENTRAL HEATING UNIT

FLOOR REGISTERS

FLOOR REGISTERS

45° ELBOW

WALL REGISTERS

ADJUSTABLE CIRCULAR ELBOW

Fig. 6.21. Typical ductwork system for Electricaire unit, showing rectangular and circular ducting components

the ducts are normally bolted to the plenum chamber on site, and are available in various sizes to suit the ducting being used.

The metal ducts are normally rectangular in cross-section, but circular types for branch-circuits can be obtained (see Fig. 6.21). Standard ducting components are laid from the plenum chamber to the required outlet positions. Flanged sections are used to attach the duct to the plenum chamber. At outlet positions, the duct can be terminated either with a 'boot' to which a floor-mounted grille can be attached, or with a right-angled bend unit which fits into the wall cavity and to which a wall-mounted grille can be attached. Special tee-joints can be employed to split the air flow into a side-duct.

The ducting components usually have one end with a slightly greater diameter than the other, and joining is normally effected by pushing one into the other and fixing with self-tapping screws or clamps and then sealing round the joint with heat-resisting tape. When the ducts have been laid in position and the terminating components fitted, the whole length of each duct is then wrapped with an insulating material, such as glass fibre, to prevent heat loss. The plenum chamber is also surrounded by insulating material and then the ducts and chamber are wrapped in a waterproof plastic or polythene sheeting overall which is taped in place. With solid-floor construction, the concrete is laid round and over the ducts. As construction of the house proceeds any rising ducts required to serve upstairs rooms are fitted, the first section being attached to the plenum chamber prior to concreting.

Some systems employ return-air ducts from the various rooms to the cupboard in which the unit is installed, and these are installed, suitably lagged, as construction of the house progresses. The grilles for return-air ducting are installed at high level, either on walls or ceilings, and the ducts taken back to the central position under floorboards or in the lofts of bungalows. Where return-air ducts are not used, the cupboard in which the unit is installed should have high- and low-level grilles, these usually being fitted in the doors.

The heating unit

This is essentially a large storage radiator, the main difference being that the casing contains a highly effective lagging which restricts case emission on the large models to about 1 kW. The lagging may be permanently attached to the inside of the heater case or inserted before the heating core is placed in position. Many different types of insulating material are used for lagging, the most common being calcium silicate slabs, often corrugated to provide airways. This is sometimes used in conjunction with a rigid insulation comprising corrugated asbestos sheets with interleaved aluminium foil.

The metal case, with the insulation in position, is bolted to the top of the plenum chamber, and the heating core is usually inserted by removal of the front of the case. The storage core consists of a number of high-grade cast iron blocks, or refractory bricks. The blocks or bricks are designed and arranged to provide grooves to accommodate the heating elements and gaps to give air passage ways. The storage capacity of each Electricaire unit is determined by the number of blocks or bricks inserted, the height of the various cabinets usually being altered to accommodate different kilowatt ratings.

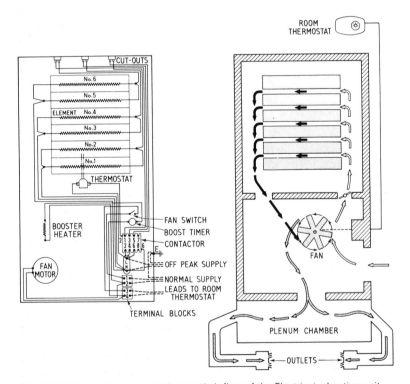

Fig. 6.22. The electrical connections and air flow of the Electricaire heating unit

The elements, either sheathed-type or 80/20 nickel-chrome wire on ceramic formers, are inserted into the grooves between the blocks. Electrical connection is made either to busbars, or interconnected porcelain-insulated connector blocks fitted inside the heater cabinet (see Fig. 6.22). Various loadings of elements are available for each size of heater.

Air is drawn into and expelled from the heater by a centrifugal fan, mounted within or on top of the heater cabinet. The fan is often driven by a separate induction motor via a belt and adjustable pulley. The heater output is determined by the fan speed and this is adjusted by alteration of the pulley to spread the amount of heat stored in the core over any given discharge period. The fan is normally controlled by a room thermostat which is installed in the most commonly used room.

The temperature of the heater core when fully charged may be of the order of 430°C, so that air blown around the heater blocks by the fan also reaches a very high temperature. The fan is, therefore, also used to mix this hot air with a regulated quantity of cold air to give an output temperature of about 60°C. The mixing of hot and cold air can be carried out in the plenum chamber or in the fan itself.

As the core cools down during discharge the temperature of the hot air coming from the core falls and a greater proportion of hot air must be mixed in to keep the outlet temperature steady. To do this, a damper or flap valve can be employed at the inlet to the storage core. This damper is controlled by a push-rod linked to a bi-metallic sensor positioned in the hot air flow from the core. This bi-metal gradually opens the damper, thus increasing the proportion of hot air to compensate for the fall in temperature. Another method is to employ a similar valve in the cold-air flow, which gradually closes as the temperature falls.

Controls

The charge controller The amount of charge is limited by using a rod-type thermostat, which projects into the core and has a control knob situated on the outside of the cabinet, thus enabling the user to select the amount of charge. Thermostats are normally rated at 15–20 A and a contactor is, therefore, necessary to switch the full heating load current of the elements.

Output boost The Electricaire specification also requires a boost in heat output to be available for short periods. There are two methods of achieving this — either increasing the fan speed or the outlet temperature of the air. In nearly all types of Electricaire unit, the boost is manually controlled, and a clockwork timer is often employed to give a time limit of up to 30 minutes.

Increasing the fan speed to give an output boost is achieved by simply switching to bring a lower resistance into the circuit. Increasing the air-outlet temperature without increased flow has the advantage of providing a boosted-heat output of about 50 per cent above normal, without any increase in noise level. The normal air-outlet temperature (usually about 60°C) is maintained by an output stabilising device. When boost output is required, a second bi-metallic device, positioned in the damper-operating rod, is activated by a small heater coil which adjusts the damper to give an air-outlet temperature of approximately 82°C for the duration of the boost period. As soon as the heater coil is de-energised the damper returns under spring pressure to its previous position and the outlet temperature reverts to normal.

Case temperature-limiting cutouts To guard against possible overheating in the event of thermostat failure, hand-reset thermal cutouts are positioned in an accessible position at the top of the heater cabinet. These cutouts are directly in series with the heater elements and do not affect the supply of the fan. They also guard against the possibility of excess case temperature due to the heater being inadvertently covered up.

Fixed wiring for Electricaire units

The electricity board will install, as required, a separate off-peak meter, timeswitch and a contactor for the Electricaire circuit. A switchfuse of 30 A or 60 A rating should be installed adjacent to the meter and suitably sized cable used for the fixed wiring, depending on the heater size. At the heater position, a double-pole switch usually of 30 A or 60 A rating must be installed to terminate the fixed heater wiring. This switch, which should preferably have a pilot-light, may be of the flush or surface type, the flush type being fitted into a steel box.

A switch must also be provided in the heater cupboard for the fan supply. This supply may be taken from any convenient unrestricted power circuit, in which case the switch must be replaced with a fused, switched connection unit. The feed to the connection unit may be looped off a nearby fitting or a junction box inserted in the circuit cable. Alternatively, the fan supply may be obtained by running a separate cable back to a way of the normal unrestricted-supply consumer unit. This is often more convenient, since one cable has already to be run to the meter position for the heater and there is little extra work involved in including a small additional cable for the fan supply. The cable for the fan fixed wiring may be 1.5 mm^2.

Both the heater and fan supplies may be terminated in a combined switch unit. Such units are available with 30 A d.p. switches for heaters up to 6 kW and d.p. 5 A fused-switches for fan supply; or with 60 A d.p. switches for heaters up to 15 kW and 5 A d.p. switches for fan supply. They are available in flush or surface type and may have flex-outlets for connection of flexible cables to heater and fan.

The flexible cables used for connection from the switch to the heater and fan must be of the heat-resisting type. Three-core flexible cables and — unless conduit systems are employed — twin-and-earth PVC-sheathed fixed wiring must be used for both heater and fan supply.

Water heaters

Generally speaking, there are three popular methods of providing domestic hot water electrically:

1   Self contained storage heaters
2   Instantaneous type
3   Immersion heaters or circulators.

Storage heaters

There are four types of storage heater, as follows:

(a) Non-pressure (known also as push-through, displacement, or open outlet); 5 to 90 litres capacity
(b) Pressure (semi-pressure) type; 20 to 30 litres capacity
(c) Cistern (incorporating a self-contained cold-water cistern); 20 to 130 litres capacity
(d) Two-in-one heater having two elements at different levels; 90 litres capacity.

Non-pressure (open-outlet) type This type of heater is used when only one point is to be supplied with hot water but it can be used to supply two adjacent points by means of a swan-neck swinging spout. The water control is on the cold-water inlet and the hot water is displaced by cold water entering the heater; hence the term 'displacement heater'. Non-pressure heaters of 14 litres capacity and under can be connected direct to a cold-water main; above this size, supply from a cold-water cistern is often insisted upon by water authorities.

Pressure-type Pressure or semi-pressure heaters are generally used for supplying a number of taps. They are always fed directly or indirectly from a separate cold-water ball-valve feed tank. The heater should be located so that

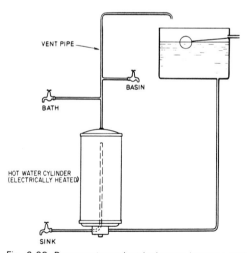

Fig. 6.23. Pressure-type electrical water heater supplying a number of taps

the shortest run of pipe is at the sink, where most of the hot water is drawn in small quantities throughout the day (see Fig. 6.23). A vent pipe must be fitted, running over the cold-water ball-valve tank. To avoid heat losses, the runs of hot-water pipes to taps should of course be kept as short as possible.
   Some models are adaptable for use as pressure or non-pressure types.

Self-contained cistern type The cistern type of water heater has the advantage that it saves the cost of extra piping and feed tanks required for other types. It is essential that this type is fixed above the highest draw-off point. It will feed any number of taps but, in order to secure efficient results, the lengths of piping feeding the taps should be kept as short as possible.

Two-in-one heaters In the dual or under-drain-board heater, advantage is taken of the fact that the full capacity of the heater is frequently not required continuously. In one particular type two heating elements, each 3 kW loading and with its own thermostat, are fitted — one near the top to keep about 30 litres of hot water available for general daily use, while the other is at the bottom, and switched on when larger quantities of hot water (up to 90 litres) are required.

The control switch is fitted on the top of the heater and permits either of the heaters to be switched on, but not both at the same time.

Instantaneous water heaters

These are of the non-pressure type and, as the name implies, they provide an immediate supply of hot water. They are particularly suitable for locations where connection to a central system would be unduly expensive or difficult and are often used to provide hot water for hand washing or showers in cloakrooms, toilets, workshops etc.

For use over sinks two models are normally available giving 1.5 or 3 litres of hot water per minute for electrical loadings of 3 kW or 6 kW respectively.

When used for showers the load is from 4 to 7 kW, and a separate circuit should be used direct from the consumer unit.

All instantaneous type water heaters include a thermostat, which is set for the water temperature required, and a safety cut-out to prevent overheating.

Some instantaneous hot water heaters for hand washing provide a continuous spray of warm water for a predetermined period — usually about 18 seconds.

Combined fuel-fired and electric system

Fig. 6.24 shows an economical way of using a pressure-type heater in series with an existing hot-water supply system. The hot water passes through the electric heater on its way to the hot taps. The electric water heater takes over completely when the fire goes out or during the summer. If it is desired at any time to use only the fuel-fired system, piping at the electric heater should be arranged with two change-over valves to allow the electric heater to be by-passed.

Electrically heated water should be prevented from circulating through towel-rails or radiators. These should be so connected that they only operate when the solid-fuel boiler is in use by originating this circuit from the flow and return pipes of the solid-fuel boiler. Electrical boilers are now available and can replace other types with little alteration to the existing system.

Immersion heaters

These are available in various types and loadings and are inserted in the hot

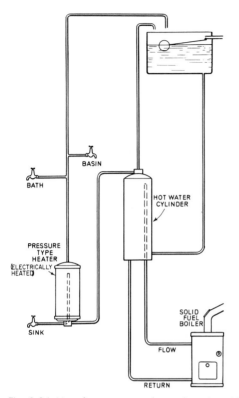

Fig. 6.24. Use of pressure-type heater in series with existing fuel fired system

water cylinder or tank. For a single heater the 3 kW size is the most usual in domestic premises.

A heater fitted horizontally towards the bottom of the cylinder will, *in time,* heat practically all the water in the cylinder, but a small amount of hot water will not be available quickly. If this is required, a further heater must be fitted at the top of the cylinder.

If an immersion heater is fitted vertically at the top of the cylinder, it will produce a small amount of hot water quickly and will heat almost all the water in the cylinder providing the heater reaches almost to the level of the cold water inlet.

'Dual' heaters are available containing two separate elements, one short and one long, in the same housing and each with its own thermostat. These heaters are fitted vertically in the top of the cylinder. The short heater is used to obtain a small amount of hot water quickly and economically, and the other when it is required to heat all the water in the cylinder.

With a dual heater or two separate heaters, the switching is usually arranged so that only one can be used at once, and special switches are available for this. However, in some cases the switching is such that either or both can be used at the same time, in these cases it must be ensured that the electrical supply circuit has adequate capacity.

Some dual heaters include an integral selector switch in the top of the

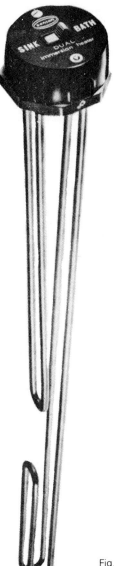

Fig. 6.25. Dual immersion heater showing selector switch
(*Santon Limited*)

heater which permits either heater, but not both, to be used at any given time. If desired this switch can be bypassed and the heaters wired to a selector switch fitted elsewhere. Fig. 6.25 shows a typical heater of this type.

Immersion heaters with a special finish are available for areas where the water is hard and aggressive since these conditions reduce the life of normal heaters. Titanium sheathing is used for very severe conditions and carries a 5 year guarantee. Where the conditions are not quite so severe an 'Incoloy' sheath is used, which is guaranteed for 3 years. Both these materials have high

Table 6.3. Immersion-heater loadings recommended for average requirements

Automatic temperature control		Hand control	
(litres)	*(kW)*	*(litres)*	*(kW)*
70	1	70	2
90–110	2	90–110	3
110–135	3	135–180	4
135–220	4	180–270	6

resistance to corrosion and scaling. In cases of this nature the manufacturers should be consulted.

Immersion heaters should normally include, or be controlled by, thermostats; otherwise boiling may occur if the user forgets to switch off.

These heaters may be connected on the normal or off-peak supply (if available) depending on the requirements. It is often useful to control an immersion heater by a timeswitch so that it is switched on and off automatically at pre-determined times; there should be an override switch so that the timeswitch can be bypassed when desired.

Table 6.3 gives the electrical loadings required for immersion heaters for varying cylinder capacities.

Insulating the cylinder or tank

The importance of careful lagging or heat-insulation of the hot-water tank or cylinder cannot be over-emphasised. Many materials are available for heat-insulation purposes. Two cheap and easily-installed materials are hair-felt and glass-fibre. Glass-fibre can be obtained in narrow boards or flexible strip form. It can be cut to fit round pipes and is easily held in position with metal bands, tape or string. Both felt and glass-fibre should be covered with canvas or other suitable material to prevent particles of hair or glass being brushed off on to clothing or other items which may be placed near the tank.

It is however preferable to use a purpose-made lagging jacket. These are easier to install and give a higher degree of heat insulation.

Thermostat settings

The thermostat control for soft water and districts up to 12° hardness may be set to cut out at a water temperature of 82°C. For hard-water areas of 12°–20° and over, the setting should not be higher than 60–66°C to reduce the rate of deposit and maintenance costs towards descaling the heater.

A storage temperature of 60°C is recommended as high enough for kitchen sink use; 66°C is scalding temperature; 43°C is the initial temperature for a hot bath. Greater storage temperatures increase standing losses, but can be used to increase storage capacity.

Immersion heaters with solid-fuel systems

Under certain conditions, the combination of a solid-fuel fire with an electric immersion heater can provide a simple and satisfactory arrangement at

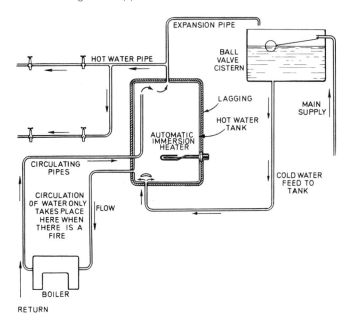

Fig. 6.26. Typical hot-water system with immersion heater

reasonable cost in new housing and for the conversion of existing installations. The solid fuel system can be shut down in the summer if desired, and the immersion heater will then provide the hot water required. Suitable conditions for using this method are:

1 When the tank or cylinder into which the heater is to be inserted is in good condition and can be efficiently lagged with heat-insulating material.
2 When the hot-water pipes feeding the taps are all taken from the vent pipe or the top of the tank; where the supply to the sink or other taps is taken from the flow pipe, such connections must be disconnected and rearranged as above.
3 When the pipe run to the kitchen sink from the hot tank is not more than 6 m (6 mm steel pipe) or 11 m (6 mm copper pipe).

Fig. 6.26 shows a typical case.

Conversion of existing cylinder

Various types of fixing flanges for converting an existing hot-water tank or cylinder to heating by immersion heater are obtainable — soldering type, bolt-on pattern, mechanical non-soldering flanges, patent flanges of many designs, either curved or flat — and the most suitable selection for any particular installation should be carefully considered. Rubber joints should be avoided if at all possible. The immersion heater should be above the cold feed to prevent heated water from circulating up the cold-feed pipe. A depth of not less than 50 mm should be provided below the cold-feed level to allow for the accumulation of sludge and scale.

When a cylinder or tank is drilled to accommodate an immersion heater, it is important to ensure that no metal particles are left inside; otherwise chemical action may occur, leading to leaks, particularly when dissimilar metals (such as copper and galvanised steel) are involved.

Electrical connections for water heating appliances

As stated earlier, it is not good practice to connect water heaters of more than say 1.5 kW loading on socket-outlet ring or radial circuits and they should preferably be on separate circuits from the consumer unit; this applies whether they are on the normal or off-peak supply.

In bathrooms, every care must be taken to ensure that the installation is in accordance with local water and electrical regulations. IEE Regulation 601-08-01 requires that no switch or other means of electrical control or adjustment is normally accessible to a person using a fixed bath or shower. If this condition cannot be met by a wall-mounted switch inside the bathroom it is necessary to fit the switch in a convenient position outside the room, or to use a cord-operated ceiling switch within the room. Wall-mounted switches should be double-pole 20 A and ceiling-mounted switches should be double-pole 15 A. Both wall and ceiling switches may be obtained with a pilot light for visual indication of 'on' and 'off'. These are well worth while.

Twin-and-earth PVC insulated and sheathed cable is normally used for the fixed wiring from the consumer unit to the local switch. When it is permissible to fit the switch adjacent to the heater, the final connection should consist of 3-core heat-resisting, flexible cable.

Where there is a considerable distance from the switch to the heater a further length of PVC-insulated cable should be taken from the switch and terminated in a wall-mounted flex outlet plate fitted adjacent to the heater.

Where a conduit system is employed, non-sheathed PVC-insulated cables may be used. Conduit which is concealed under the plaster should be terminated at a metal box, using a bush and locknut or other terminating method. A flex outlet plate may then be fitted on the box to enable the final connection to be made with heat-resisting flexible cable. The earth core in the flexible cable to the heater should be connected to the earth terminal in the conduit box.

Where surface-mounted conduit is employed, it may be terminated in a surface-mounted metal box or, if the design allows, it may be taken to the terminal chamber of the heater. In each case, a bush and locknut or other suitable terminating device should be used. If the conduit is connected directly to the terminal chamber of the heater, a bonding clip must be attached to the conduit, after thorough cleaning, and a bonding lead of the required size taken to the earth terminal of the heater. The earthing leads should be identifiable by green and yellow insulation or tape.

A water-pipe is not acceptable as the sole means of earthing, and should not be used as a protective conductor. Earthed metalwork should, however, be bonded to any adjacent water pipes or metalwork in accordance with IEE Regulations. Shower units should be on a separate circuit and controlled by a 30 or 45A DP cord-operated ceiling switch, with pilot light.

7 Survey of modern wiring systems

A general survey of the chief methods of electric wiring should be useful to the installation engineer in that it will help him to decide which method is most suited to any given conditions. Cost, reliability and ease of erection are the main factors to be considered. When quoting for installation work, it is often advisable to give the customer a choice of two systems, with a brief statement on the advantages and disadvantages of each. Separate chapters are devoted to the more important systems, the installation of conduit systems being dealt with in Chapter 8, and other systems in Chapter 9.

Steel conduit systems

In steel conduit systems, steel tubes are fixed to the walls, etc., and cables, usually single-core PVC-insulated, non-sheathed, are drawn into them afterwards. Although the tubes can be fixed to the surface of walls, etc., the full advantages of this method are realised in a building which is in the course of erection, as the conduit may be fixed to the unplastered walls, or chased into the brickwork, and then completely covered with plaster. If the cables are not drawn in until the plaster is quite dry, they are protected from any moisture which might leak into the tubes from the wet plaster.

Types and grades of conduit

Conduit is supplied in standard lengths of about 4 m and is manufactured in accordance with BS 31 or 4568 and is supplied with both ends of each standard length threaded; if shorter lengths of conduit are required it can be cut and the ends re-threaded on site by using a screwing machine or stock-and-die set.

Conduits are screwed together by means of threaded sockets, tees, elbows, etc. Conduit tubes merely slipped together and held only by self-tapping screws, or left unlocked in any way, are not suitable for wiring and are not permitted — IEE Regulations 543-03-03 refers. When completed, the whole conduit system must be electrically continuous, so that a current applied to any part of the conduit will be conducted to the main earthing terminal for the installation.

Conduit is either made of heavy-gauge steel strip formed into a tube with the seam welded, or solid drawn. The latter is very much more expensive than welded to produce and its use is virtually restricted to gas- or vapour-proof installations.

Oval conduit

Light-gauge conduit is also obtainable in oval form. This is particularly suitable for use where the depth of plaster is insufficient to allow for the standard round conduit without chasing the walls. Oval tubing is available in close-jointed and brazed grades. It is more expensive than the corresponding round tubes. Suitable fittings for making conduit runs in 'oval-to-oval' and 'round-to-oval' tubing are available, thus permitting easy connecting between both types of conduit on the one installation, if so desired.

Waterproof installation

When a waterproof-conduit installation is required, the only way to obtain it is by means of screwed joints and watertight fittings; it must, however, be remembered that all screwed fittings are not necessarily watertight and when required, they must be distinctly specified.

Galvanised, sheradised and enamelled conduit

All conduits and respective fittings are available in the following finishes:
    Galvanised — mainly for external use.
    Sheradised — for external use, or internal use in wet or damp situations
    Enamelled — for internal use in dry situations.
Some conduits are also available in a light silver-grey finish for internal use where surface systems are to be employed. They can easily be painted over if required, or remain unobtrusive if left unpainted.

Choice and erection of conduit

Heavy-gauge screwed conduit is considered to be the soundest system. It affords complete protection of the cables and the screwed joint is more secure from the point of view of making the conduit electrically and mechanically continuous.

The recommended method of installation, whether with screwed or continuity-grip fittings, is for the conduit to be erected in position as a rigid entity before any cables are installed. This complies with IEE Regulation 522-08-02.

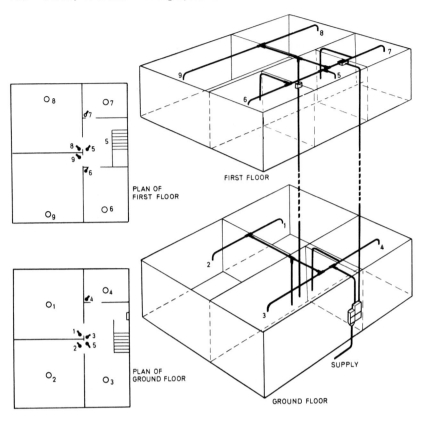

Fig. 7.1. Layout of conduit for electric lighting in a typical semi-detached house

The main runs are first taken from the plans of the installation and locations for junction and draw-in boxes selected, from which branch circuits will radiate to the various points served by that main run. One branch run may be arranged to carry the cables which supply several switch positions near together. The same method can be adopted for several adjacent lighting points. Draw-in boxes are included in the runs as necessary to facilitate the drawing-in of the cables to supply the various points (see Figs. 7.1 and 7.2).

A common mistake to be avoided, whatever the method of erection, is the overcrowding of cables in the conduit. This may lead to excessive strain when drawing cables into the tube and may fracture some of the copper strands, thus lowering the current-carrying capacity of the cable. The insulation may also be damaged and the cable rendered useless. In a well-planned installation, it should always be possible to withdraw from a conduit any wiring that may have proved faulty without disturbing the rest of the wires in the same conduit. Outgoing and return cables carrying a.c. should always be installed in the same conduit; this requires particular attention when wiring 2-way switches. The erection of steel conduit and the cable capacities of conduits are dealt with in the next chapter.

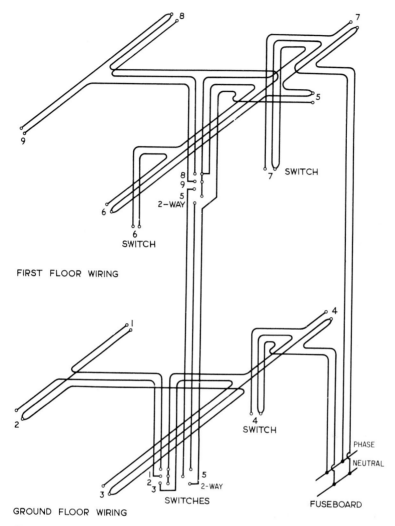

Fig. 7.2. Cables in a conduit lighting installation. Where cable is looped into switch or light positions the conduit carries at least three cables

Non-metallic conduit systems

While steel conduits provide excellent protection against mechanical damage and give earth-continuity when initially installed, condensation, rusting and corrosion may present problems. There is also the possibility of electrical continuity being impaired by corrosion during the life of the installation.

Non-metallic conduits can be obtained which:

(a) are highly resistant to corrosion by water, acids, alkalis and oxidising agents in the concentrations likely to occur in most buildings, and are unaffected by concrete or plaster;

(b) are dimensionally stable and non-ageing;
(c) do not deteriorate significantly after long external exposure;
(d) will not support combustion;
(e) are not susceptible to water condensation; and
(f) have excellent electrical properties (breakdown voltage is often 12–20 kV/mm.

There are several manufacturers producing non-metallic conduit. Most employ PVC (polyvinyl-chloride) compounds for conduit and fittings. Two distinct types are available — rigid and flexible.

Rigid conduit

The most commonly used type of rigid conduit is made of 'unplasticised' PVC in accordance with BS 4607, and is obtainable in outside diameters of 16, 20, 25, 32, 38 and 50 mm. It is normally supplied in 4 m lengths with plain bored ends and is available in two wall thicknesses — that is, either heavy (standard) gauge or light gauge.

Heavy-gauge conduit is normally installed on surface installations where there is some risk of mechanical damage. It can, if required, be threaded by using normal stocks and dies. Threading of PVC conduit is not, however, often specified since this weakens the wall thickness and creates what is termed 'notch sensitivity'.

Light-gauge conduit is always used with plain fittings and cemented at joints if necessary. This type of conduit is normally used for concealed work, although some manufacturers produce a white, light-gauge conduit with matching fittings that can be used on surface installations where the risk of mechanical damage is slight.

Flexible plastic conduit

Flexible plastic conduit, available in coiled lengths of 25 m, is used for sunk or concealed wiring, where appearance is of no significance. Its flexibility is an advantage, as it allows awkward bends to be negotiated and threading through holes in such structures as bison floors. The conduit readily adapts itself to irregularities in wall surfaces, etc., and it can be used for wiring movable equipment. It can withstand the stresses and strains imposed by barrows and boots while awaiting covering with floor screeds, etc.

Flexible conduits of differing sizes are usually designed to sleeve over each other and short pieces of the appropriate sizes can therefore be used as coupling sleeves. Thus, to join two 20 mm conduits together, a 50 mm length of 25 mm flexible conduit (20 mm i.d.) can be cut and sleeved over the joining conduits, sealing with a solvent or suitable compound for strength and watertightness. Adhesives are available from the conduit manufacturers.

Rigid and flexible oval conduit

Oval conduit is often used for switch-drops and socket-outlet drops since it does not require deep chasing and is not affected by plaster. Special wide-oval conduit can be obtained to accommodate two 2.5 mm² twin-and-earth cables side by side.

Fittings

A full range of fittings and accessories is available for use with insulated conduit, catering for all conditions likely to be met. The range includes couplings, saddles, adaptors, bushes, tees, bends, and junction boxes of all types. Square and rectangular boxes which accommodate standard lighting switches or socket-outlets are also available.

Effect of temperature

Rigid PVC conduit and fittings are not suitable for use where the ambient temperature is likely to fall below −5°C, or where their normal working temperatures exceeds 70°C. If installation conditions present border-line temperatures, the manufacturers should be consulted, since in some instances where the heat is only intermittent the problem can be overcome. In low temperatures, the conduit becomes harder and less ductile and should therefore not be installed in positions where it is likely to be subjected to continuous blows, e.g. from opening doors, etc.

Where a PVC outlet box is used for the suspension of a lighting fitting, care must be taken to ensure that the temperature of the box does not exceed 60°C. Suspension of totally enclosed lighting fittings from standard PVC boxes should be avoided unless special metal fixing clips are used (Fig. 7.3). The weight suspended from a PVC box should not exceed 3 kg unless metal supports are inserted.

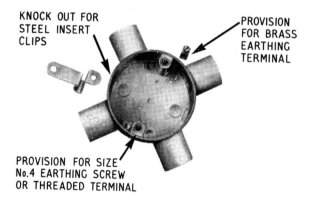

KNOCK OUT FOR STEEL INSERT CLIPS

PROVISION FOR BRASS EARTHING TERMINAL

PROVISION FOR SIZE No.4 EARTHING SCREW OR THREADED TERMINAL

Fig. 7.3. A circular PVC junction box with spouts for conduit entry. Steel insert clips can be used to provide heat conduction or support for heavy pendant fittings (*Egatube Limited*)

The linear coefficient of thermal expansion for PVC is 6 to 8×10^{-5} per °C. This is roughly equivalent to an expansion of 12 mm in a 4 m length for a temperature rise of 45°C. For this reason, expansion couplers should be used where rigid conduit is installed in straight runs (i.e. no bends or sets) for lengths in excess of 5.5–6.5 m. The saddles which are used to support PVC conduit should also allow lateral movement.

For details of installation methods see Chapter 8.

All-insulated sheathed wiring systems

One type of cable is used in all-insulated systems: PVC (polyvinylchloride) insulated and sheathed cable; it is manufactured as single-core (known as 'singles'), two-core (known as 'twin'), and two-core with a bare protective conductor between them (known as 'twin-and-earth'). A three-core cable is also available which has three insulated cores within the sheathing.

Sheathed wiring cables are used for surface-wiring systems, or are simply buried beneath the plaster or concrete, although mechanical protection may be necessary in particular instances. PVC-sheathed cables will resist attack by most oils, solvents, acids and alkalis. They are not greatly affected by the action of direct sunlight and are non-flammable. For these reasons PVC-sheathed cables are suitable for a wide range of internal and external applications.

PVC-sheathed cables may be run between floors and ceilings and dropped down through ceilings to socket-outlets, switch points, etc. Holes made for the passage of cables through ceilings should be filled in with cement or other suitable material as a precaution against the spread of fire.

Fig. 7.4. Circular 30 A junction box for use with PVC sheathed cables (*Ashley Accessories*)

PVC-sheathed cables running along walls may be buried direct in the plaster but it is preferable, and involves very little extra cost, to protect them by means of metal or PVC conduit or channelling. This gives some protection against nails driven into the wall and, in the case of conduit, permits the cables to be withdrawn or replaced at a later date.

Accessories used with PVC-sheathed cables are usually of the insulated type; however metal boxes can be used at outlet points if desired, provided suitable grommets or brass bushes are fitted where the cables enter the boxes. Fig. 7.4 shows a typical all-insulated junction box; if a metal box is used, suitable screw terminals should be fixed inside the box.

For full installation techniques see Chapter 9.

Cabling to accessories and fittings

All-insulated wiring systems with PVC sheathed cables are used for most domestic lighting and power applications.

Twin-and-earth, PVC-insulated (or polythene-insulated) PVC-sheathed cables are normally used for both lighting and power since the protective conductor must be taken to all lighting points and switches as well as to socket-outlets.

Where the cable is concealed beneath the plaster, it must be terminated in a box. The box may be the plaster-depth type, which is screwed to the brickwork of the wall, or a standard or deep box which may require some chasing of the brickwork.

At lighting-point positions, it is usual to take the cables straight into the ceiling rose or pattress block; which forms the incombustible enclosure required by IEE Regulation 526-03-02. At socket-outlet, switch or lighting-point positions, the box or accessory must contain all the unsheathed portions of cable; this also applies to any junction boxes used.

Where cabling to socket-outlets, switches or lighting points is run on the surface, a box need not be used at these positions provided that the accessory or fitting has an enclosure into which the unsheathed portion of cable may be taken. This enclosure must be of incombustible material and may be an integral part of the fitting, or may be formed by a part of the fitting and an incombustible part of the building structure. This does not preclude the use of boxes for surface wiring systems.

Many flush fittings may be used with surface wiring in conjunction with a special moulded box for surface mounting. Whether metal or insulated boxes are used for switches in insulated wiring systems, an earthing terminal must be provided on them or on the switchplate and connected to the protective conductor. The connection of wiring to socket-outlets, ceiling roses and switches is covered in Chapters 4 and 5.

Mineral-insulated copper-sheathed wiring system

Mineral-insulated copper-sheathed (MICS) cable consists of solid copper conductors insulated with a highly compressed covering of magnesium oxide

(MgO) powder and sheathed with a seamless malleable-copper tube. The main advantages of MICS cable are:

Robust — does not need any further mechanical protection except in areas where damage is likely.

Non-flammable — will withstand very high temperatures and fire (copper will withstand 1,000°C).

Relative position of the conductors cannot be disturbed by bending or twisting.

Impervious to oil and water and not affected by condensation, provided the ends are properly sealed.

Use of inorganic materials means that cables are non-ageing and if properly installed should last indefinitely.

Cables are comparatively small in diameter for the current-carrying capacity and are easily bent — the minimum bending radius is six times the overall diameter.

The copper sheath provides excellent earth continuity.

The cable is sufficiently pliable to be used for final connection of cookers and other normally fixed apparatus that occasionally have to be moved for maintenance or cleaning; a short loop should be left in the cable in such cases. MICS cables may be run along the surface of walls and structures and fastened by copper clips and saddles which are supplied to fit the overall cable diameter and number of cables. Ferrous clips or saddles should not be used with bare MICS cables owing to the risk of chemical action, particularly under damp conditions.

Table 7.1. Spacing of supports for MICS cables

Overall diameter of cable (mm)	Maximum spacing of clips	
	Horizontal (mm)	Vertical (mm)
Not exceeding 9	600	800
Exceeding 9 and not exceeding 15	900	1200
Exceeding 15 and not exceeding 20	1500	2000

Note: The spacings stated for horizontal runs may be applied also to runs at an angle of more than 30° from the vertical. For runs at an angle of 30° or less from the vertical, the vertical spacings should be applied.

These cables may be laid under floors or buried in concrete, plaster or in the ground. The IEE Regulations require that MICS cables exposed to the weather or risk of corrosion or laid underground or in concrete ducts must have an overall PVC sheath. In addition, underground cables require to be protected by cable covers or indicated by marker tape (IEE Regulation 522-06-03).

A PVC sheath overall gives some protection against mechanical damage but in some cases cables running on the surface at low level may require to be enclosed in conduit or channelling. The PVC sheath is available in a variety of colours for circuit identification, but black or orange are the standard colours.

Table 7.2.

Current-carrying capacities (amperes):

Ambient temperature: 30°C
Sheath operating temperature: 70°C

Nominal cross-sectional area of conductor (mm²) 1	2 single-core cables or 1 two-core cable, single-phase a.c. or d.c. (A) 2	3 single-core cables in trefoil or 1 three-core cable, three phase a.c. (A) 3	3 single-core cables in flat formation, three-phase a.c. (A) 4	1 four-core cable 3 cores loaded three-phase a.c. (A) 5	1 four-core cable all cores loaded (A) 6	1 seven-core cable all cores loaded (A) 7	1 twelve-core cable all cores loaded (A) 8	1 nineteen-core cable all cores loaded (A) 9
Light duty 500 V								
1	18.5	15	17	15	13	10	—	—
1.5	23	19	21	19.5	16.5	13	—	—
2.5	31	26	29	26	22	17.5	—	—
4	40	35	38	—	—	—	—	—
Heavy duty 750 V								
1	19.5	16	18	16.5	14.5	11.5	9.5	8.5
1.5	25	21	23	21	18	14.5	12.0	10.0
2.5	34	28	31	28	25	19.5	16.0	—
4	45	37	41	37	32	26	—	—
6	57	48	52	47	41	—	—	—
10	77	65	70	64	55	—	—	—
16	102	86	92	85	72	—	—	—
25	133	112	120	110	94	—	—	—
35	163	137	147	—	—	—	—	—
50	202	169	181	—	—	—	—	—
70	247	207	221	—	—	—	—	—
95	296	249	264	—	—	—	—	—
120	340	286	303	—	—	—	—	—
150	388	327	346	—	—	—	—	—
185	440	371	392	—	—	—	—	—
240	514	434	457	—	—	—	—	—

Notes: 1 For single-core cables, the sheaths of the circuit are assumed to be connected together at both ends
2. For bare cables exposed to touch, the tabulated values should be multiplied by 0.9

Table 7.3.
Volt drops for single-phase operation (mV/A/m)

Ambient temperature: 30°C
Sheath operating temperature: 70°C

Nominal cross-sectional area of conductors (mm²)	Two single-core cables			Multicore cables		
	$mV/A/m_r$	Touching		$mV/A/m_r$	$mV/A/m_x$	$mV/A/m_z$
	$mV/A/m_r$	$mV/A/m_x$	$mV/A/m_z$			
1	2	3	4	5	6	7
1	42	—	—	42	—	—
1.5	28	—	—	28	—	—
2.5	17	—	—	17	—	—
4	10	—	—	10	—	—
6	7	—	—	7	—	—
10	4.2	—	—	4.2	—	—
16	2.6	—	—	2.6	—	—
25	1.65	0.200	1.65	1.65	0.145	1.65
35	1.20	0.195	1.20	—	—	—
50	0.89	0.185	0.91	—	—	—
70	0.62	0.180	0.64	—	—	—
95	0.46	0.175	0.49	—	—	—
120	0.37	0.170	0.41	—	—	—
150	0.30	0.170	0.34	—	—	—
185	0.25	0.165	0.29	—	—	—
240	0.190	0.160	0.25	—	—	—

Current ratings

MICS cables have higher current ratings than other cables, e.g. PVC or rubber-insulated, of the same conductor size. Conversely, for a given current rating, a MICS cable will usually have a smaller conductor size and overall diameter. This is due to the high thermal conductivity of the magnesium oxide insulation (which allows the heat generated in the cable conductors to flow quickly from the cable) and the higher permissible operating temperatures. However it is still necessary to check the voltage drop and in many cases this, rather than the current rating, determines the cable size, particularly when long runs are involved.

It will be seen from Table 5.3 (page 64) that for ring and radial circuits feeding socket-outlets to BS 1363 the conductor sizes for MICS cables are less than for other types of cable.

Table 7.2 gives the current ratings for certain MICS cables. Ratings for other types are given in Appendix 4 of the IEE Regulations.

MICS cable is available in two grades, i.e. light duty (600 V) and heavy duty (1000 V). Light duty cable is normally quite adequate for domestic installations. Installation of MICS cables is covered in Chapter 9.

Choice of system

Of the three main systems, i.e. conduit, PVC-sheathed and MICS cables, the PVC-sheathed system is the cheapest and is the most widely used for domestic installations.

Conduit systems or MICS cable can of course be used on domestic work but they are normally used only on high-class installations where cost is not the primary consideration.

8 Installation of conduit systems

As explained in Chapter 7, there are two main conduit systems, the metallic (usually steel) and the non-metallic (plastic). As might be expected, the methods of installation of these are quite different and they are dealt with separately in this chapter.

Steel conduit

Although steel conduit is the premier method of encasing wires, a great deal depends upon the quality and manner of assembly.

Before commencing any work on conduit systems, the size, type and number of cables that are to be drawn into each conduit must be decided, as this will influence the size of conduit and fittings. Almost any low-voltage cable can be drawn into conduit. The most commonly used are PVC-insulated cables or rubber-insulated braided and compounded cables but the latter are rarely used now.

IEE Regulation 521-02-01 requires that cables of a.c. circuits installed in steel conduit must always be so bunched that the cables of all phases and the neutral conductor (if any) are contained in the same conduit. This requirement also applies to the 'go' and 'return' cables in a circuit, even if the neutral is not involved, e.g. the two wires to a one-way single-pole lighting switch.

Cables installed in conduit are normally PVC-insulated single-core, non-sheathed type to BS 6004. These are available in a single voltage range and are suitable for single-phase (240 V) or 3-phase (415 V between phases and 240 V between phase and neutral). Where extra protection against damp or mechanical damage is required, PVC-*sheathed* cable may be drawn into conduit. Owing to their greater diameter, the size of conduit used must then be carefully considered.

Size of conduit

When the size and type of cables for the various circuits has been decided and the runs worked out, the correct size of conduit to accommodate these cables must be selected.

Tables giving the number of cables that may be drawn into the various sizes of conduit are given in Table 8.1.

Table 8.1. Capacities of conduits (metallic or non-metallic). PVC-insulated single-core, non-sheathed cables, straight runs not exceeding 3m in length

(a) Cable factors for short straight runs

Type of conductor	Conductor cross-sectional area (mm²)	Factor
Solid	1	22
	1.5	27
	2.5	39
Stranded	1.5	31
	2.5	43
	4	58
	6	88
	10	146

(b) Conduit factors for short straight runs

Conduit dia. (mm)	Factor
16	290
20	460
25	800
32	1400

(Based on IEE Regulations)

The tables apply to metallic or non-metallic conduits, but a separate protective conductor is required with the non-metallic type.

Other tables cover respectively (a) straight conduit runs not exceeding 3 m in length, (b) straight runs exceeding 3 m in length, and (c) runs of any length incorporating bends or sets. For each cable conductor size a factor is given and the factors for all the cables which are to go in the same conduit are added together. Each conduit size is also given a factor and a conduit is chosen having a factor equal to or higher than the total of the cable factors. The resultant number of cables should not be exceeded.

If it is desired to allow spare capacity for additional cables in the future these must be allowed for in addition to the initial requirements.

For situations not covered by these tables, advice should be obtained from the manufacturers.

Screwed fittings

To enable continuous runs of conduit to follow curves and varied paths and also to overcome the necessity of separate runs of conduit for each circuit, various fittings such as elbows, joint-boxes, switch-boxes, etc., can be used. These are supplied with their connecting-points female threaded with the standard tapping, to allow the threaded conduit to be screwed into them. Their types and shapes are extremely varied due to the desire of manufacturers to produce fittings requiring a minimum of labour yet giving a maximum of efficiency. They can, however, be classed under 'inspection' and 'non-inspection' types (see Fig. 8.1).

NORMAL BEND
(SCREWED)

INSPECTION BEND
(SCREWED)

Fig. 8.1. Conduit fittings

Inspection-type fittings have a removable lid, or cover, held in place by two or more small screws which fit into tapped holes. They should be installed so that they will remain accessible for such purposes as the withdrawal of existing cables or installing additional cables.

Non-inspection (solid) conduit elbows or tees should be restricted to the following positions:

1   Locations at the ends of conduits immediately behind a lighting fitting, outlet box or accessory of the inspection type.
2   One solid elbow located at a position not more than 500 mm from a readily accessible outlet box in a run of conduit not exceeding 10 m between two outlet points, provided that all other bends in the run are not more than the equivalent of one right angle.

A reduction in the number of cables drawn into conduit must be made when non-inspection elbows and tees are used.

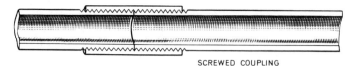

SCREWED COUPLING

Fig. 8.2. Conduit coupler

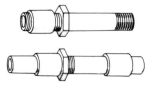

Fig. 8.3. Running joint

Where greater lengths of conduit are required than those supplied, couplers are employed (Fig. 8.2). A running joint is used where it is impossible to turn screwed conduit (see Fig. 8.3).

Conduit boxes

The chief use of these boxes is to facilitate drawing in wires by giving access to them at sharp bends (Fig. 8.4). The internal diameter (60.3 mm) renders them unsightly for surface work for which the smaller inspection-type fittings, such as elbows, tees and bends, are more suitable. On sunk and concealed work, however, circular boxes are to be preferred. They are made in a number of patterns — two-way 'through', two-'angle', three-way, four-way, back outlet, etc. (Fig. 8.5).

The covers are of varying patterns, usually of sheet iron, and held in position by two small screws. Sometimes the outer rim of the cover is flush with the outer rim of the box. This type is quite satisfactory when the boxes are used under floors but when sunk in plaster, it is better to use a break-ring or a cover of slightly larger diameter than the box; this allows the joint

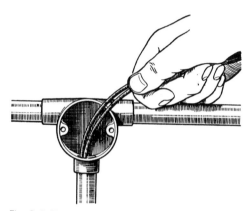

Fig. 8.4. Three-way conduit box. The large aperture makes for easy working in positions such as under floors where neatness is not an essential factor

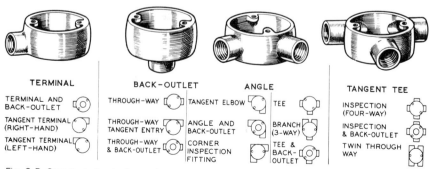

Fig. 8.5. Standard circular boxes for use as draw-in boxes or for mounting accessories at 51 mm fixing centres

between the plaster and the box to be effectively concealed. In place of iron, fibre covers are sometimes used; these enable fittings such as switches or ceiling roses to be screwed to them.

Another type of cover is made of cast iron and is thicker than the sheet iron variety. It has one face machined smooth. The round box used with this type has its rim machined also, and when in position the joint is practically watertight. As an extra precaution a rubber or fibre gasket may be inserted between the surfaces.

Other types of boxes, employed solely for housing switches, socket-outlets, connection units, etc., are more usually made of pressed steel, with a non-corrosive finish (see Fig. 8.6). Knockouts are provided in the sides and back for conduits and tapped lugs are fitted to the sides for attaching the fitting with metal-threaded screws (see 'Metal boxes for accessories' on page 125).

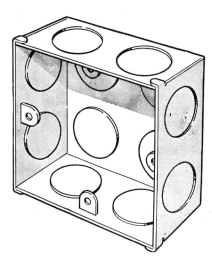

Fig. 8.6. Pressed steel conduit box showing knockout sections and fixing lugs for accessories

Fig. 8.7 shows two methods of achieving the same objective on a conduit system. It is a matter of opinion which is preferable. Circular boxes are not a necessity.

Rectangular boxes are made with outlets in all the usual positions, while for unusual situations, undrilled boxes can be supplied, and the outlets drilled and tapped by the operative, as required.

In place of conduit boxes, tees and bends of the 'inspection' type may be used, although in practice these are only considered when neatness and restricted space are governing factors. The size of the 'inspection' aperture is partly determined by the bore of the conduit and the available working space is necessarily small.

Fixing looping boxes

Where the conduit is to be concealed in floors or walls, it is often convenient to use looping boxes at which the cables can be drawn in. Details of the

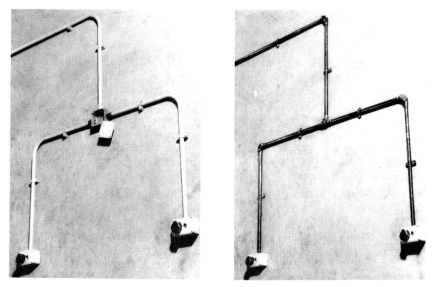

Fig. 8.7. Conduit layouts (left) the conduit box has been 'set' and a BS box used; (right) elbows and tees have been used

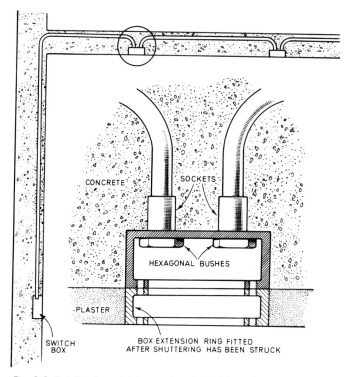

CONCRETE

SOCKETS

HEXAGONAL BUSHES

PLASTER

SWITCH
BOX

BOX EXTENSION RING FITTED
AFTER SHUTTERING HAS BEEN STRUCK

Fig. 8.8. Details of conduit box and method of fastening conduit. Note that the bushes should be firmly tightened

looping-box system are shown in Fig. 8.8. Looping boxes are available with two to four holes at the back, by which the conduit can be teed off to other points as required.

The conduit may be fixed to the box by means of a socket at the back and hexagon brass bush inside, or by means of a locknut and a brass tube end. A collar extension piece or ring on the box enables it to be deepened as required to be level with the plaster surface.

To fit the looping boxes shown in Fig. 8.8, special tools have been designed and these are illustrated in Fig. 8.9. The hexagon bush may be tightened with

Fig. 8.9. Special tools for installing looping boxes; (left) ring bush securing tool; (centre) cutting tool for clearing enamel from holes in conduit boxes; (right) double ended spanner

an ordinary spanner, but it may be found more convenient to use the double-ended spanner shown on the right of the illustration. This spanner can be fitted over the end of a bush and enables a very tight connection to be made with the end of the conduit.

The expanding tool shown on the left of Fig. 8.9 may be used for connecting the looping box using the brass-tube end and lock-nut method. The connection is made in the following manner. The lock-nut is screwed on to the threaded end of the conduit which is then inserted into the hole of the looping box. The brass-tube end is next placed over the expansion end of the tool and the tommy bar rotated. This causes the end of the tool to expand, thus firmly gripping the tube end by its internal periphery whilst it is being screwed on to the conduit. When the tube end is securely tightened on to the conduit, rotation of the tommy bar in the reverse direction causes the expanded end to return to normal, so releasing the tube end.

The third tool shown in Fig. 8.9 has a special cutting edge and may be used for clearing enamel from the holes of looping boxes.

Metal boxes for accessories

An essential fitting for all conduit work is the metal box for fitting accessories such as socket-outlets, switches, etc. In addition to forming a suitable housing for the accessory for protective purposes, this is also necessary to complete the protective conductor path of the system.

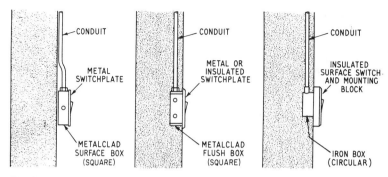

Fig. 8.10. Three methods of terminating conduit at a metal box

Varying circumstances call for different types of boxes (Fig. 8.10). For sunk work, flush boxes of round or square pattern can be obtained. The number and positions of entries differ according to requirements, as back entry only, back and side entry, or side entry only. Metal boxes for concealed conduit work are normally constructed of pressed steel and have tapped lugs attached to the sides for fixing the flush accessory. They usually have a selection of 16 mm and 20 mm knockouts and the fixing lugs are tapped at 60.3 mm fixing centres.

The type of box used for flush switches in sunk work should not be employed in surface installations. A seamless type of surface box for flush switches is obtainable. Where two or more switch plates are required to be assembled in one position 'multiple' boxes effect a great saving in labour and materials, with the added advantage of neatness — the description of types for single boxes being also applicable to these. However, 2-gang switches are available mounted on a single square switch plate.

Fig. 8.11. Weatherproof switches for use with surface conduit

For watertight installations a special switch and box complete is made, watertightness being effected by fitting the switch cover with a rubber gasket and the operating handle rotating in a gland or bush (see Fig. 8.11).

Methods of fixing conduit

It is usually best to commence fixing the conduit at the farthest point of the installation and to work towards the supply point or distribution board. The type of fixing varies according to the method of installing the conduit. Where screwed tubing is to be placed on the surface in exposed positions, this should be done with spacer bar saddles (see Fig. 8.13). Some examples are illustrated, but there are others with wood or iron fixing. If sound fixing to brickwork is required, a saddle with the shank formed as a rag bolt for building-in can be obtained. The important feature of spacer bar saddles is that the conduit is held clear of the wall, and water or moisture drains away instead of being held to cause corrosion. Spacer bar or distance saddles permit cleaning to be carried out behind the conduit, which is important in certain areas.

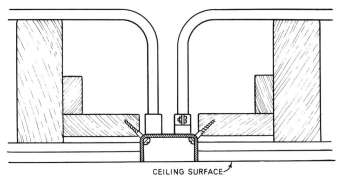

CEILING SURFACE

Fig. 8.12. Lug grip looping box for ceiling point, showing a method of securing the box by means of screws through knockouts at the back

Internally, ordinary saddles as illustrated form a good fixing. The half saddle while not being so rigid, has the advantage that the conduit can be held close to the angle of a ceiling or wall. The second lug on the other type would prevent such close fixing. Saddles are also supplied in longer varieties known as multiple saddles, these being useful for holding two or more conduits at one fixing.

Conduit which runs beneath floors should be fixed to the joists with metal saddles. If the run of the conduit is across the joists, they must be slotted at each point of crossing. The slots must be of the minimum possible size and kept as near as possible to the bearings of the joists. These requirements are extremely important and must be met even though the length of the conduit run is thereby increased. Excessive cutting of the joists may seriously weaken the building.

When the tubing is buried in plaster or chased into brick walls, the plaster to some extent holds it in position, and in such cases it can be fixed with crampets or pipe-hooks (see Fig. 8.13). These are driven into the joints of the brickwork at intervals, so that the conduit is held firmly to the wall. The conduit should be covered by at least 6 mm of plaster in sunk work. On surface work, pipe-hooks are unsightly and the little trouble expended in fixing saddles is well worth while.

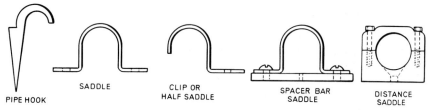

Fig. 8.13. Fixing for metal conduit

Nailing is not satisfactory for fixing surface conduit, and on tiles or brickwork where a good wood-screw fixing cannot be obtained, the most satisfactory method is to use wall plugs. The spacing of supports for rigid metal conduit is given in Table 8.2.

Table 8.2. Spacing the supports for rigid metal conduits

Nominal conduit size (mm)	Maximum distance between supports	
	Horizontal (m)	Vertical (m)
Not exceeding 16	0.75	1.0
Exceeding 16, not exceeding 25	1.75	2.0
Exceeding 25, not exceeding 40	2.0	2.25
Exceeding 40	2.25	2.5

(Based on IEE Regulations)

Conduit assembly

The differences between faulty and successful workmanship usually depend on small matters. The first essential of all conduit work is that it shall be properly prepared. The ends should be rounded off with a file until they are smooth — rough edges may cause abrasions in the cables. Conduit is generally cut with a hacksaw; this leaves sharp burrs on the edge of the tube, which must be completely filed away. The interiors of all fittings should be examined for any rough finish.

The ends of conduit, however smooth they are made, should always be terminated with proper bushes if threaded boxes are not used. There are two or three types on the market, brass being used for screwed conduit. For the cheaper forms of installation, rubber or fibre bushes will be found quite satisfactory.

Cleanliness is another factor in good assembly work. Paint, enamel and rust are poor conductors; therefore if a good earthing system is to be obtained, all conduits and fittings must have the ends well cleaned down to the bare metal before the joints are made. A loose joint also means poor contact and a loss of earthing efficiency; see that nuts or screws are properly tightened and screwed ends tightly threaded.

Above all, bear in mind that electrical continuity throughout is very important. Avoid the use of insulated boxes where possible, unless at the ends

Fig. 8.14. A serrated conduit coupling for use with a smooth-bore brass bush (*Walsall Conduits*)

of branches. A PVC junction box, for example, breaks continuity, and it is much easier to use a metal box than to connect effectively the conduits feeding a PVC box.

Where connections are made to boxes that have enamelled finishes, for example, the enamel must be removed at the point of conduit entry. The serrated conduit coupling shown in Fig. 8.14 will greatly assist in maintaining good continuity at switch-boxes, etc.

Where conduits are terminated in switchfuses or metalclad consumer units an earth terminal is usually provided on the side of the case for connection of the earthing conductors. Where satisfactory connection cannot be made through the casing, earthing clips must be attached to each conduit and a bonding lead connected between them and the main earthing terminal. IEE Regulation 543-02-07 requires that at all socket-outlet positions, an earth terminal must be incorporated in the outlet box and a separate protective conductor connected from this to the earthing terminal of the socket-outlet. This bonding lead should be insulated and coloured green and yellow.

On screwed conduit work it may be found necessary to cut a thread on a length of conduit; a common error is to make a longer thread than necessary for the required connection. The conduit is enamelled or galvanised for protection and this over-cutting leaves a portion of exposed metal which should be painted with a non-corrosive paint after the joint has been made if future trouble is to be avoided.

Making a system watertight

When installing a watertight system, all joints should be painted with a metallic paint, covers of boxes should be properly seated and where there is any doubt there should be no hesitation in fitting gaskets under covers. If it is unavoidably necessary to make a joint in the wiring, the added precaution should be taken of filling the box with a waterproof compound.

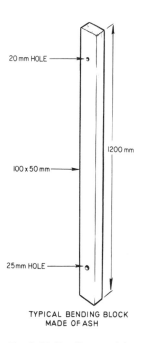

20 mm HOLE

1200 mm

100 x 50 mm

25mm HOLE

TYPICAL BENDING BLOCK
MADE OF ASH

Fig. 8.15. Bending conduit using a bending block

Fig. 8.16. Bending machine for medium sized conduit

Bending and setting conduit

Conduit runs often pass from one direction to another not at right angles to the first, or small changes of level may be necessary. When surface conduit is used in conjunction with iron-clad switchgear and distribution boards, a small set-off of the conduit is required to make a secure mechanical and electrical contact with the equipment. In these cases, the conduit must be bent.

A bending block or machine is used for bending and making set-offs in conduit of larger sizes, such as 20 mm, 25 mm and above (see Figs. 8.15 and 8.19). A bending block consists of a small baulk of timber with holes drilled at the ends slightly larger than the sizes of the tube to be bent. If the tube is pushed through a hole and the other end of the timber is placed on the ground, sufficient leverage on the tube can be obtained to make quite a satisfactory bend. A mechanical bender is a more efficient device.

Feeding in the cables

With inspection fittings, spaced at appropriate intervals throughout the installation, the cables can usually be pushed in from one inspection fitting to the next. Some specifications require that wires be pushed in and not pulled, the reason being that it is impossible to push more wires into a tube than it will comfortably hold. However, the draw-wire or draw-tape, if properly used, can be an invaluable implement, and there are many cases of erection where it can be used without detriment to the job, and where its use is undoubtedly a time-saver.

If installation of the cables is commenced from the mid-point of the conduit run, the length of cable being drawn is minimised. The cables should be fed in slowly and steadily and in such a manner that they do not become twisted or crossed (see Fig. 8.17). Any chafing against the boxes should be avoided.

Fig. 8.17. Drawing the wires into the conduit at a junction box. The wires are kept parallel by allowing them to slide through the fingers

The draw-wire or draw-tape is a wire or steel tape at the end of which there are either two little wheels or runners set at right angles to each other, or merely a brass ball. This end is inserted into the empty tube and is fed in by hand. The front end of the wire or tape is guided round all the bends and angles in the run until it emerges either at the other end of the tube or at one of the inspection fittings that have been provided.

If a tape is used it should not be used for pulling in the cables. A draw-wire should be attached to the tape and pulled through first. When it emerges, the cables to be drawn into the tube are fastened to the tail end of the draw-wire and pulled in. There is often a temptation to save money on tubes by employing a draw-tape and cramming each tube with more cables than it should hold. This practice must be avoided at all costs.

Cooke's cutters

Cooke's cutters should form a part of an electrician's equipment. This is an implement for cutting a hole in a steel or iron plate where ordinary drilling is out of the question. In using this tool, a small pilot hole is first drilled in the plate, box, tank or other object, after which a centre pin is passed through the hole thus drilled from back to front and the cutter engaged with this pin by means of a thread. The cutter is screwed down by means of a mill-headed screw until it bears upon the surface of the plate and is then rotated with a tommy bar, being continually fed down on to the job with the mill-headed screw.

Non-metallic conduit

Non-metallic (PVC) conduits and fittings were described in the previous chapter. There are several manufacturers of PVC conduit and fittings, but the following instructions on installation may be taken as a general guide.

Saddles or clips used for fixing PVC conduits should be of the type which allows longitudinal movement of the conduit to cater for expansion. Saddles or clips should be spaced according to Table 8.3, but where conduits are

Table 8.3. Spacings of supports for non-metallic conduits

Nominal conduit size (mm)	Maximum distance between supports			
	Rigid		Flexible	
	Horiz. (m)	Vert. (m)	Horiz. (m)	Vert. (m)
Not exceeding 16	0.75	1.0	0.3	0.5
Exceeding 16, not exceeding 25	1.5	1.75	0.4	0.6
Exceeding 25, not exceeding 40	1.75	2.0	0.6	0.8
Exceeding 40	2.0	2.0	0.8	1.0

(Based on IEE Regulations, Table IIC)

installed in high-ambient-temperature conditions or where rapid temperature changes occur, these fixing centres should be suitably reduced. Where 'sets' are made, fixings should be made at 150 mm on each side of the bends. Non-metallic conduits need not be drilled for drainage as very little internal condensation occurs.

Bending of rigid conduit

Cold bending of rigid PVC conduits up to about 25 mm outside diameter can be quite successfully carried out for most of the angles encountered in normal domestic installations. PVC has a tendency to straighten out after bending and it is therefore necessary to overset each bend to allow for this. In cold weather, it is often helpful to rub briskly over the area where the bend is to be made with a rag to create sufficient friction to warm the conduit and make bending easier.

Standard bending machines can be used to bend and set PVC conduit but it is more usual to employ a bending spring. This is a spiral spring which is inserted into the conduit and manoeuvred into the position where the bend is to be made by using a draw-wire. The spring has an eye on one end to facilitate this, and also to allow the spring to be turned in an anti-clockwise direction which will make it easier to withdraw in the case of very tight bends. The conduit should not be bent back with the spring inside.

For larger sizes of conduit, i.e. 32 mm, 38 mm and 50 mm and also for smaller conduits in cold weather, heat should be applied when bending. This can be carried out as follows. First, warm the conduit over an electric fire, spirit stove, butane lamp or yellow flame of a blowlamp. The conduit should be heated for at least 300 mm on each side of the bend. After about one minute of continuous movement within the flame or heating area, the conduit should become soft and flexible. It may then be placed on the floor and bent to the desired shape or a bending core can be inserted to prevent distortion. The conduit should be held in the bent position until it hardens and becomes rigid again, which will normally take approximately one minute.

Coupling rigid conduit

At junction boxes, fittings or switch and socket boxes, coupling can be readily carried out by using plain bore (push fit) types with a solvent adhesive for additional strength or watertightness. Conduit fittings made by one manufacturer have special sockets and locking rings to hold the conduits in place. Screwed-entry fittings can be obtained, if required, but are not recommended for two reasons: electrical continuity is not involved and therefore a threaded section is unnecessary, and by threading conduits, the wall is weakened and what is termed notch-sensitivity is created.

However, if threading of conduit is required, this can be easily carried out by using standard stocks and dies (the handles of which should be removed for easier working) and holding the conduit in one hand and the stock in the other hand. New dies should always be used and kept separate for use on plastic conduit. Lubrication is not normally required, but with very long threads, Vaseline or soft soap may be used.

Where more than two lengths (about 8 m) of conduit are run on the surface, or where temperature variations of 14°C or more are likely, expansion couplers should be fitted. These are special sockets with extended entries to allow the conduit to move up and down as it expands. The conduits should first be pushed right into the coupler and then withdrawn approximately 10 mm. If it is necessary to make the joints watertight, a non-hardening compound should be used; on no account should a solvent be applied.

Flexible conduit

Flexible conduit can be used (a) where flexibility is required in surface installations, such as motor connections, and (b) in concealed work where appearances do not matter or where obstacles or other difficulties prevent the use of rigid conduit. Typical situations are under concrete screeds, threading under floorboards or other restricted spaces, threading through ceilings or negotiating awkward bends. Flexible conduit can also be used to provide bends in an installation using rigid PVC conduit. Instead of bending the rigid conduit, a length of flexible conduit can be cut, formed by hand into the required bend and slipped over the two ends of the rigid conduit. Provided the rigid conduit is firmly held by saddles on each side of the bend, the flexible conduit will retain its formed shape.

Flexible conduits of differing sizes are designed to fit over each other, and short pieces of the appropriate sizes can thus be used for coupling, as previously indicated in Chapter 7.

Connection to metal boxes

Non-metallic (PVC) conduits may be used with standard cast-iron or pressed-steel accessory boxes. Connection is made by use of PVC adaptors which may

Fig. 8.18. Flexible non-metallic conduit drop to switch point run in wall angle to avoid wall chasing. Conduit is held in position with cement

be clip-in plain bore, or screwed at one end and plain bore entry at the other end. The screwed types may be male-threaded for use with locking rings or female-threaded for use with male bushes.

An earthing terminal must be provided on all metal boxes and plastic switch boxes and connected to the protective conductor.

Wiring non-metallic conduits

Cables used are the PVC-insulated non-sheathed type. Where it is necessary for a draw wire to be used for 'pulling-in', a liberal smear of grease or soft soap at the entering position, or a smear of liquid paraffin on the cables themselves, will assist greatly.

The maximum number of cables (see Table 8.1) should be adhered to — remember that a separate protective conductor must be included, and connected to the main earthing terminal at the supply position. An earthing terminal must be provided at all lighting switch and outlet positions connected to the protective conductor of the final circuit. Switch boxes and ceiling roses with a moulded-in earth terminal can be obtained for this purpose. The protective conductor must be insulated and identifiable at all outlet, switch or junction positions by green and yellow insulation sleeving.

9 Installation of PVC-sheathed and MICS cable

As indicated in Chapter 7, the main alternatives to conduit systems are (a) PVC-insulated and sheathed cables, and (b) MICS cables. These may be run on the surface or concealed.

Surface wiring is normally restricted to existing property or other areas where it would be inconvenient or expensive to conceal the cables or where appearance is not important.

PVC-sheathed wiring

This is the most common system for domestic installations. Where the cables may be subject to mechanical damage they should be protected by channelling or conduit. The protective conductor must be taken to all lighting switches and lighting outlet points, in addition to socket-outlets, and hence the cables should be twin-and-earth, or three single-core sheathed cables must be used. 3-core sheathed cables with earth conductor are also available.

In planning an installation, the number and sizes of cables must first be determined and the most suitable routes considered. Having given these our attention, we begin at the farthermost point from the consumer unit, in exactly the same manner as if we were running the wires in conduit.

The cables should be fixed at intervals, as given in Table 9.1, with special

Table 9.1. Spacing of cable supports for non-armoured PVC or v.r.i. insulated cables in accessible positions

Overall diameter of cable (mm)	Horizontal (mm)	Vertical (mm)
Not exceeding 9	250	400
Exceeding 9 and not exceeding 15	300	400
Exceeding 15 and not exceeding 20	350	450
Exceeding 20 and not exceeding 40	400	550

Note: Where cables are not of circular cross-section, the diameter is taken as the measurement of the major axis

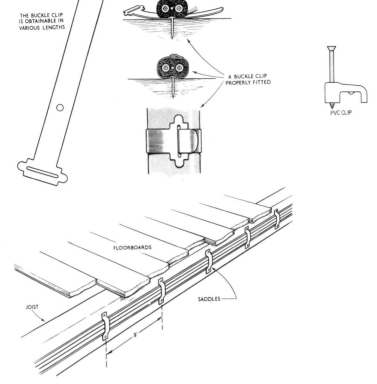

THE BUCKLE CLIP
IS OBTAINABLE IN
VARIOUS LENGTHS

A BUCKLE CLIP
PROPERLY FITTED

PVC CLIP

FLOORBOARDS

JOIST

SADDLES

9″

Fig. 9.1. Clips and saddles for sheathed wiring

clips or saddles. In dry situations, tinned-brass buckle clips may be used (see Fig. 9.1). In damp situations or where they are exposed to the weather, non-corrosive clips and brass nails or screws should be used. An alternative method, now widely used and suitable for all situations, is a PVC-moulded clip with single-hole fixing. The internal surface of the saddle is formed to suit the size of cable.

Where cables are installed in normally inaccessible positions and are resting on a reasonably smooth horizontal surface no fixing is necessary. On vertical runs in inaccessible positions, fixing is not necessary on runs of up to 5 m in length but must be provided on longer runs.

Table 9.1 gives the requirements for supporting cables.

Terminating at switches and socket-outlets

A box is not necessary at outlet or switch positions for surface-mounted items, provided the accessory or fitting has an incombustible chamber or such a chamber can be formed by the fitting and the building surface. It is, however, usual to fit surface-mounted socket-outlets and switches on to insulated pattress blocks (see Fig. 9.2).

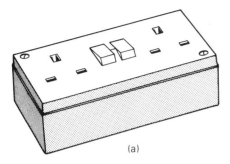

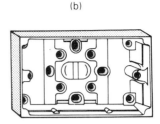

Fig. 9.2. (a) Surface mounted twin socket with mounting block
(b) Moulded box for surface mounting of flush type twin socket
(*MK Electric Limited*)

Alternatively, by using a shallow box for switches and a deeper box for socket-outlets, the same fittings that are normally used for flush mounting can be installed on the surface. These boxes are screwed to the wall and the fittings attached by metal threaded screws into tapped fixing lugs on the box (see Fig. 9.2).

Where the sheathing of the cable is cut back to make the connection to an accessory or fitting, all unsheathed portions of cable must be completely enclosed. Where junction boxes are used, all unused cable entries should be blanked off.

Surface-mounted accessories have knockout sections in the side walls for entry of cables. Where these knockouts are removed, any sharp edges should be removed with a file.

Moulded-insulated accessories and fittings should be used with surface-wiring systems and all switches and ceiling roses should be fitted with an earth terminal. Where composite twin-and-earth cables are used, the bare protective conductor should be covered with a green-and-yellow coloured insulating sleeve or insulating tape at switch and ceiling-rose positions. Fig. 3.9 shows a plastic box used to convert a flush-fitting switch to surface-mounting, and the insulating sleeve fitted to the protective conductor.

Running cable

Where the cable passes through walls or ceilings, the holes must be made good with an incombustible material to prevent the spread of fire. A short length of bushed conduit or sleeving should also be used to prevent any sharp edges on the brickwork from chafing the sheathing. The tubing should also be filled with incombustible material.

Cables running under wooden floors or above ceilings should be installed in such positions that they are not likely to be damaged by contact with the floor or ceiling or their fixings. Cables which cross wooden joists should pass through holes drilled in the joists so that the cable is at least 50 mm measured vertically from the top or bottom of the joist as appropriate (see Fig. 9.3). If this is impracticable, the cables should be protected by enclosure in earthed

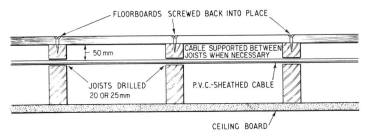

Fig. 9.3. Installing PVC-sheathed cable under floorboards

metallic conduit or equivalent mechanical protection (IEE Regulation 522-06-05). If the spacing between joists exceeds the value in Table 9.1 for the cables concerned, the cables must be supported by suitable wooden battens, or by other means.

The floorboards should preferably be screwed (rather than nailed) back into position and, in any case, a screwed section of board should be placed over each junction box or other position where access may be required at a later date.

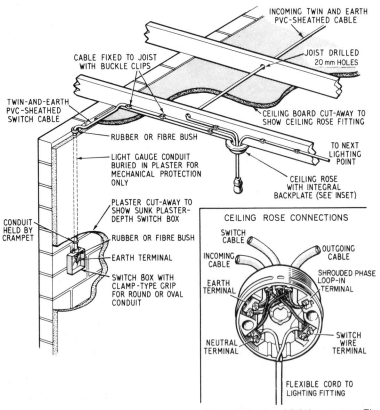

Fig. 9.4. Use of four-terminal ceiling roses with an all-insulated lighting system. The inset shows the connections of a ceiling rose

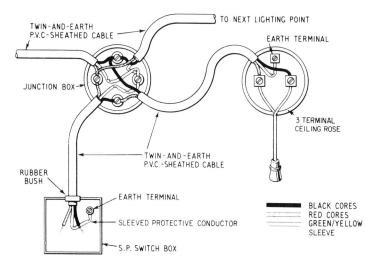

TO NEXT LIGHTING POINT

TWIN-AND-EARTH
P.V.C-SHEATHED CABLE

EARTH TERMINAL

JUNCTION BOX

3 TERMINAL
CEILING ROSE

TWIN-AND-EARTH
P.V.C.-SHEATHED CABLE

RUBBER
BUSH

EARTH TERMINAL

SLEEVED PROTECTIVE CONDUCTOR

S.P. SWITCH BOX

BLACK CORES
RED CORES
GREEN/YELLOW
SLEEVE

Fig. 9.5. Use of junction boxes in an all-insulated looped lighting circuit

Where cables pass through structural steelwork, the holes must be fitted with suitable bushes to prevent abrasion of the cables.

Fig. 9.4 shows the cabling of a lighting circuit using 4-terminal ceiling roses, and Fig. 9.5 shows the equivalent arrangement using junction boxes. For more information see Chapters 4 and 5.

Installation hints

When pulling PVC-sheathed cables off drums, the cable should not be allowed to 'spiral' off. It is easy to arrange for a metal rod to be supported in such a manner that the drum may be slipped over it and allowed to revolve as the cable is drawn off.

Care must be taken when removing the sheath of PVC cables to ensure that the conductor insulation is not damaged.

The amount of sheathing removed for connection of cables into accessories or fittings should be just sufficient to enable the conductors to be connected without undue tension, and such that when the accessory or fitting is screwed into position there is not too much slack conductor at the back.

When baring the ends of the conductors for connection, care should again be taken to ensure that the conductor strands are not cut or nicked, as this may eventually mean a reduction in current-carrying capacity.

All conductor clamping screws should be well tightened, as loose connections produce high-resistance points which can cause sparking with the associated danger of insulation charring.

Cable runs should be planned so that crossing of cables running on the surface is avoided as far as possible. During installation, the cables should be pulled tight and flat against the wall or ceiling and the fixing clips lightly tapped with a hammer to keep the cables in place.

Mineral-insulated copper-sheathed cables

MICS cables can be fixed to the surface with saddles or clips, or if required can be plastered or concreted over. Bending can be done by hand or by using special tools. The cable-bending radius should, however, be limited to six times the diameter of the cable; this allows the cable to be straightened out at a later date without ill-effect, if required (see Fig. 9.6).

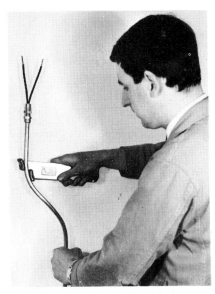

Fig. 9.6. Bending and setting MICS cable. The rubbing surfaces of the tools are faced with leather to avoid damaging the cable sheath

The cable ends need not be sealed during installation, since moisture penetrates very slowly and the affected magnesia will fall away when 'stripping back' to provide tails. However, when the sheath has been stripped back, the seal should be made immediately.

Sealing cable ends

A complete termination comprises two sub-assemblies, each performing a different function — the seal (which excludes moisture from the cable) and the gland (which is used to connect or anchor the cable at a box). The seal consists of a brass pot with a disc, to close the mouth, and sleeves to insulate the conductor tails. Compound is used to fill the pot.

There are two types of seal, standard and medium temperature. Fitting both types is identical and the same tools are employed. The standard seal comprises a brass pot, phenolic disc and PVC insulating sleeves (see Fig. 9.7). It is suitable for continuous operating temperatures up to 105°C. The medium temperature seal comprises a brass pot, glass-fibre disc and f.e.p.

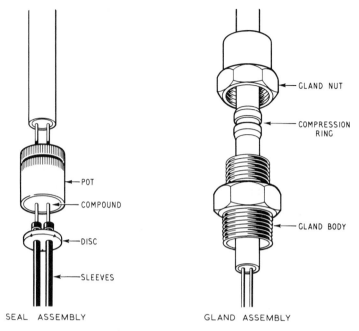

Fig. 9.7. Seal and gland assemblies for MICS cable

(fluorinated ethylene propylene) sleeves. It is suitable for continuous operating temperatures up to 150°C.

The black PVC sleeves for the standard seal are supplied in 300 mm lengths with a moulded head at each end. The sleeves may be cut to provide 150 mm per conductor, to a length appropriate to the conductor tails required. The cut lengths are threaded through the holes in the disc and retained by the

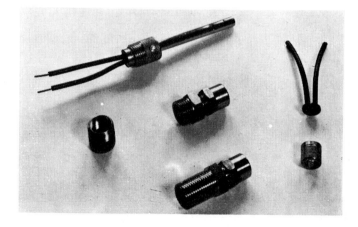

Fig. 9.8. MICS cable terminations; (top) earthing screw type gland for cable entering grip conduit box; (centre) short and long threaded glands; and (right) cap and sleeving sub-assembly and screw-on pot

moulded heads. Some manufacturers provide plain lengths of sleeving and a wedge must be fitted in the end of each sleeve to form an 'anchoring bulge'.

Two types of gland can be obtained: the standard or compression type employing a ring, or olive, which is compressed into the sheath of the cable (Fig. 9.7); and the earthing-screw type, employing two screws through the barrel of the gland which are tightened on to the cable sheath. The standard type is the most reliable and efficient method, but the earthing-screw type is quite adequate for domestic installations and is much cheaper (Fig. 9.8).

When choosing the type of gland required, the entry thread for insertion into the box or fitting must be borne in mind and also the bore of the gland must be appropriate to the size of cable being used. Standard glands are available with British electric thread in sizes 12, 20, 25 and 32 mm, the thread length normally being 16 mm in each case. The earthing-screw gland is only available for the smaller sizes of cables used in domestic installations and thread size is either 16 mm or 20 mm.

Preparing the cable end

When preparing the cable ends for sealing, the following sequence should be observed:

1   Cut the cable to the length required, allowing for the appropriate length of conductor tails. The cable should be cut off squarely to facilitate stripping. Where the MICS cable has a PVC sheath overall, the PVC should be cut back before the copper sheath is stripped off.
2   Mark the point to which the copper sheath is to be stripped back to expose the conductors. The sheath should be indented by using a special ringing tool which will cut a small groove, thus ensuring that the sheath breaks away cleanly in the correct place. The correct depth for the groove is half the depth of the sheath and to obtain this, tighten the wing-nut on the tool until the cutting wheel is firmly in contact with the sheath, then give an additional quarter to half a turn according to the cable size. Rotate the cutter around the cable for one complete turn, or more if the first cut is not quite deep enough.
3   Remove the copper sheath, using one of the three methods mentioned below, but make sure that the cable end is square and free from burrs after stripping.
4   Shake off any loose magnesia and thoroughly clean the conductors.

The cable end is now ready for applying the seal, and this should be done as soon as possible to avoid the absorption of moisture into the insulant. It must also be remembered to thread any gland parts on to the cable at this stage.

Removing the copper sheath

There are three methods of removing the copper sheath of MICS cable by hand: using side-cutting pliers, using a fork-ended stripping rod, or using a rotary stripping tool.

The use of side-cutting pliers has the advantage of employing a tool normally carried in any electrician's tool-kit. The sheath is indented at the required point using a sheath-cutting tool as described earlier. Using side-cutting pliers, make a small tear in the end of the sheath. This tear can then be firmly gripped and by twisting the pliers around the sheath it is easy to remove the sheath in a spiral. The pliers should be kept at an angle of about 45° to the line of the cable, until the indented ring is approached, when the pliers should be brought to right-angles with the cable. The final spiral should be taken off by using the points of the pliers held parallel to the cable.

The fork-ended stripping rod is used in a similar manner to side-cutting pliers. The cable sheath is indented and a small tear made as above. The small tear is fitted into the fork end of the rod and the rod is revolved at an angle around the sheath until the indented ring is approached, when the rod should be positioned parallel to the sheath.

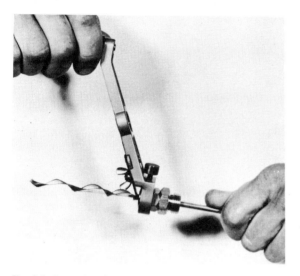

Fig. 9.9. Removing the sheath from MICS cable with a rotary stripping tool. The tool also has a wrench for screwing on the pot (*BICC Limited*)

When a rotary stripping tool is used there is no need to indent the cable or make a tear to start. A gland (minus the compression ring) of suitable size for the cable being stripped is fixed into an appropriate hole on the tool. The cutting blade is set according to manufacturer's instructions, which is usually just below the inner edge of the sheath, but not fouling the conductors. The cable end is then inserted into the gland and pressure is applied while the tool is revolved in a clockwise direction so that the blade engages with the sheath. The handle of the tool can then be firmly gripped and continued revolving will remove the sheath in a continuous spiral (see Fig. 9.9). A clean square end can be obtained by gripping the sheath with a pair of pliers and allowing the tool to revolve up against them.

Fitting the seal

Before fitting the seal, any shrouds, hoods or gland parts should be threaded on to the cable.

The brass pot should be cleaned before use and any metal particles removed. The pot has self-cutting threads and by pushing it squarely on the cable it can be engaged finger-tight. A pair of pliers or pipe grips can then be used to screw the pot down until either the sheath lip is level with the shoulder inside the pot or the threads begin to bind. If a gland is being used, a check should be made from time to time during screwing to ensure that the pot is going on square and will fit in its normal position in the gland.

An alternative method of screwing the pot in position is to use a pot-wrench tool (see Fig. 9.10). The pot is positioned between the gland and the tool and the gland are screwed finger-tight into the tool. The whole assembly is then pushed, gland first, over the conductors and the pot is screwed on to the sheath as above. The tool can be freed by unscrewing the gland, taking care not to disturb the pot. Any loose powder must now be removed by shaking or tapping the sheath.

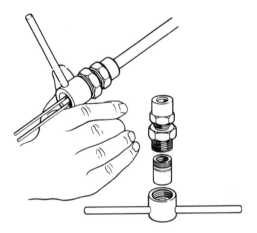

Fig. 9.10. Using a pot-wrench tool

The sleeves and disc that have already been assembled can now be slipped over the conductors with the moulded heads or beaded ends nearest the pot, and a check made to ascertain that the disc fits the top of the pot. The disc assembly should then be partially withdrawn along the conductors to allow the pot to be filled with compound. The compound is pressed into the pot using the thumb, which must be clean of any metal particles (see Fig. 9.11). The pot should be filled from one side only to avoid trapping air bubbles, and slightly overfilled so that a small mound is formed beyond the mouth of the pot.

The disc and sleeve assembly should then be pressed into the pot, using finger-and-thumb pressure to remove any surplus compound. The disc is forced into and held in position by use of a crimping tool. If the disc is notched, these should be lined up with the crimping pins on the tool. A slow, steady pressure should be applied to eject any surplus compound and air pockets without damage. The crimping tool should be screwed until the top plate is level with the top of the pot.

The cable cores must be identified by coloured tape, sleeves or discs (see IEE Regulation 514-06).

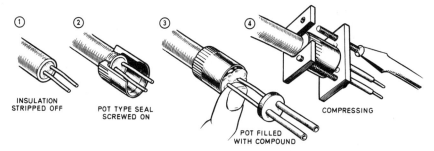

Fig. 9.11. Cold sealing for MICS cable. Having ringed and stripped the copper sheath as in (1), screw on pot (2); test the cap and sleeving sub-assembly for fit before packing plastic compound into pot (3); then force home the cap and sleeving sub-assembly with compressing tool

Where a gland is being used, this can now be brought into position, together with the compression ring and screwed home to complete the assembly.

Bonding

A range of glands, lock-nuts, reducing nipples and sockets is available for terminating the cables at any standard boxes or casings designed to take steel conduit of either the screwed or grip type. These glands, which are slipped on

Fig. 9.12. Plaster-depth switch box for MICS cables with attached clamp to grip the sheath (*MK Electric Limited*)

to the cables before the seals are fitted, firmly anchor the cables and provide an effective earth-bonding system. For surface wiring to accessories mounted directly to the wall, the cable may be terminated without the use of a conduit gland by using special boxes employing a clamp which grips the cable sheath, not the pot (see Fig. 9.12).

General notes on installation

At the final stage of manufacture, MICS cable is formed on drums or in coils of a specified diameter, hence the length of one turn of cable on a drum or in a coil can be worked out. Estimation of length can therefore be fairly accurately made by counting the number of turns and part-turns.

When pulled from a coil or off a drum the cable often forms 'waves'. These can be effectively removed by the use of a roller straightener, which is clamped over the cable and run backwards and forwards by hand with gradual progression along the cable. Cables can be straight-jointed by using special sleeves and two glanded terminations or by fixed screw terminals fitted in a metallic adaptable box. Where cable runs in excess of standard length coils are required, cables can be joined by factory made joints if required. Where cables are used in mildly corrosive conditions, spacer-bar or distance saddles should be used to lift the cable clear of the corrosive surface. Alternatively, PVC-oversheathed cable may be used. As stated in Chapter 7, MICS cables exposed to the weather or risk of corrosion or laid underground or in concrete ducts should have an overall PVC sheath.

For neatness when the cables have been installed, they may be dressed with a wood block and hammer.

When offsetting the cable to enter a fitting, a bold offset should be made, with several inches of straight cable between the set and the gland. This looks neater than a sudden offset adjacent to the fitting and also allows the gland to be easily withdrawn from the fitting if required.

Each length of MICS cable should be tested for insulation resistance between conductors and between conductors and sheath immediately after sealing and the test reading should be infinity.

PVC-insulated, PVC-sheathed cables

The installation of these cables follows normal practice, as explained in Chapter 9.

Rising mains

Rising mains will not normally be encountered on domestic work, the only exception being in multi-storey buildings such as blocks of flats. In such cases the consumers' meters are normally located on the individual floors, and rising mains are used to carry the bulk supply up the building. Such mains may consist of one of the following systems:

1   Rigid conductors in a protective enclosure
2   Single or multi-core paper-insulated or PVC-insulated and sheathed cables or MICS cables. These are run on cleats or a cable tray in a vertical chase which should be free from combustible material. Multi-core paper or PVC-insulated cables are normally armoured.
3   PVC-insulated, single-core, non-sheathed cables enclosed in conduit or trunking.

Where cables are installed in vertical enclosures, internal barriers are required at not more than 5 m centres to prevent excessive temperature rise at the top of the enclosure.

Fire barriers are also necessary at all floor levels in accordance with IEE Regulation 527-02-01. The enclosure should be adequately ventilated as necessary, particularly at the top.

The connections to the rising mains for the individual sub-circuits are made at each floor, the actual method depending on the system employed for the rising mains and the method of sub-distribution.

Busbar system

Rising main busbar systems consist of bare or lightly insulated copper or aluminium busbars contained in a metal enclosure. The busbars may be of circular or rectangular cross-section.

It is essential that the system allows for the longitudinal expansion of the busbars when loaded. This means that the bars must be free to move in their fixings; the bars are usually solidly clamped at the bottom of the run. On long vertical runs, flexible expansion joints may be necessary at intermediate points. If adequate provision for expansion is not made, serious 'drumming' or other difficulties may occur when the bars are loaded, particularly if the enclosure consists of ferrous material. Some manufacturers use non-ferrous material, e.g. brass, for the front cover of the enclosure.

The above aspects should, of course, be catered for by the manufacturer of the system but the installer should be aware of the possible difficulties.

When cables are connected to busbars it is necessary to guard against excess temperature in the cable (IEE Regulation 523-03-01).

10 Safe and efficient earthing

The necessity for earthing non-current-carrying metalwork of wiring systems and exposed metalwork of electrical apparatus derives from the fact that the neutral point of most low voltage distribution systems is earthed, so that a voltage exists between earth and the other pole or poles (Fig. 1.4).

IEE Regulation 130-09-01 requires that, where metalwork of equipment (other than current-carrying conductors) may become live due to a fault and this condition would cause danger, the following precautions must be taken:

1   the metalwork should be earthed in such a manner as will cause discharge of electrical energy without danger, or
2   other equally effective precautions must be taken.

The precautions referred to in (2) above are covered in Sections 411 and 413 of the Regulations. These measures are aimed at preventing contact with any parts which are live at normal mains voltage. The requirements in (1) above for earthing of the metalwork are detailed below.

It should be noted that a contact by a person or livestock with exposed metalwork which is not normally live but is made live by a fault, is called an 'indirect contact'. Contact with a part which is normally live is called a 'direct contact'.

There are certain exemptions to the requirement that metalwork must be earthed, and these are detailed in IEE Regulation 471-13-04. It should be noted that it is not necessary to earth short lengths of metal conduit for the mechanical protection of cables and certain other metal enclosures providing mechanical protection for equipment.

When metalwork is to be earthed in accordance with the above, the circuits concerned must be protected by either:

1   an overcurrent protective device, i.e. normally a fuse or overcurrent circuit-breaker, or
2   an earth-leakage circuit-breaker of the residual-current-operated type.

Method 2, i.e. an ELCB, must be used when the prospective current on an earth fault is insufficient to operate the overcurrent protective device; this situation arises when the impedance of the path to earth for fault currents is excessive.

However ELCBs are now more widely used, even when not strictly necessary in accordance with the above, so as to provide extra protection. ELCBs are described in detail later in this chapter (pages 168–170).

Definition of earthing terms

The earthing of metal parts is achieved by connecting them to a 'protective conductor' which is connected to the 'consumer's main earthing terminal'. This in turn is connected via the 'earthing conductor' to the 'earth electrode'. Definitions of these terms are given below.

Protective conductor   The conductor connecting to the consumer's main earthing terminal, or to each other, those parts of an installation which require to be earthed. It may be, in whole or in part, the metal conduit, trunking or duct, or the metal sheath or armour of a cable or the special conductor of a composite cable or flexible cord, or a separate conductor.

Consumer's main earthing terminal   The terminal or block of terminals to which all protective conductors are connected before connection to the earth electrode. Normally situated adjacent to the consumer's main supply terminals.

Earthing conductor   The final conductor by which the connection from the main earthing terminal to the earth electrode, or other means of earthing, is made.

Earth electrode   A conductor or group of conductors in intimate contact with, and providing an electrical connection to, earth. The use of pipes of public gas or water supply undertakings as the sole earth electrode is not, however, permitted (see IEE Regulations 542-02).

IEE Regulations, Part 2, shows these items in diagram form.

Earth fault loop impedance (EFLI)   This is defined as the impedance of the earth fault current loop, or phase to earth loop, starting and ending at the fault. It consists of the following parts:

1   the protective conductor from the fault to the consumer's main earthing terminal
2   the consumer's main earthing terminal and earthing conductor
3   the return path to the transformer neutral; this (assuming the system is not PME) may consist of either or both of (i) the general mass of earth or (ii) the metal sheath or armour of the supply cable. On a PME system the return path to the neutral point on the transformer is via the neutral conductor
4   the transforming winding
5   the phase conductor from the transformer to the fault.

The impedance of the fault is usually assumed to be negligible. Fig. 10.1

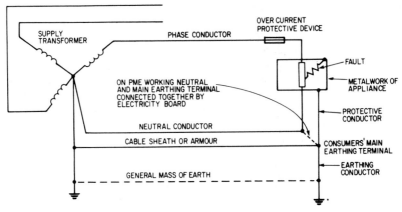

Fig. 10.1. Earth-fault loop impedance path

shows the above in diagram form. The maximum permissible values of EFLI are given in IEE Section 413 for varying types and ratings of overcurrent protection devices.

The impedance values for *socket-outlet* circuits protected by fuses given in IEE Table 41B1 are based on the supply being disconnected under fault conditions within 0.4 seconds. The values in IEE Table 41D for circuits supplying *fixed equipment* are based on a maximum disconnection time of 5 seconds. The longer disconnection time for fixed equipment than for socket-outlets is based on the assumption that there is less risk of a prolonged contact under a fault condition with fixed equipment than with portable appliances fed from socket-outlets. Values for MCBs are given in Table 41B2.

If the maximum values for socket-outlets quoted in Table 41B1 cannot be met, it is permissible to use Table 41C of the Regulations. This method is based on a maximum disconnection time of 5 seconds, as for fixed equipment. However, if this method is used the impedance of the *protective conductor*, i.e. the earth path from the socket-outlet to the consumer's main earthing terminal, must be limited to the value given in Table 41C in addition to the overall maximum value of *EFLI* as stated above. The object of limiting the impedance of the *protective conductor* is to ensure that the touch voltage on metallic parts under a fault condition does not exceed 50 V.

Bonding

The metal casings of all the electrical apparatus in an installation should be bonded together and to earth. Metal conduit should be made electrically continuous by means of screwed conduit and fittings. The sheaths of metal-sheathed cable, where employed, should also be bonded together at all junctions.

Where cables are protected by an insulated sheath or insulated conduit, earthing may be provided by a bare conductor within the cable sheath or a separate protective conductor run inside the conduit connected at the consuming end to the metal casing of the appliance, or the earth terminal of a socket-outlet.

In addition to the metalwork of electrical equipment or cables as indicated above, it is necessary to connect to the main earthing terminal any metallic main water or gas pipes and any other service pipes or ducting which enter the building; also any risers or ducts for central heating or air-conditioning systems and any exposed metallic parts of the building structure — IEE Regulation 413-02-02.

The bonding connections to main water or gas pipes should be as near as possible to the points of entry into the building but, if there is an insulating section at that point, the connection should be on the consumer's side of it. For a gas service pipe the connection should be on the consumer's side of the meter, between the meter and any branch pipework, and preferably within 600 mm of the meter. These requirements are detailed in the IEE Regulations, Section 547, which also specifies the sizes for bonding conductors. There are also special requirements for bonding to metalwork of other services on PME systems — see later in this chapter.

In case of doubt regarding the method of connection to metalwork of another service or the presence of an insulating section, the authority responsible for that service should be consulted.

In bathrooms or rooms containing a fixed shower, special conditions apply with regard to the bonding of exposed metalwork as already indicated in Chapter 6 (page 86). This subject is dealt with more fully later under 'Special precautions in bathrooms' (page 162).

Obtaining earth continuity

Earth continuity may be obtained in a number of ways throughout an installation, and some of the methods that may be employed with conduits, wiring systems, socket-outlets and plugs are illustrated.

Fig. 10.2 shows the method of ensuring earth continuity with screwed steel conduit. With screwed conduit it is essential that the threads engage tightly. Fig. 10.3 shows an earthing clamp suitable for bonding to water pipes. Fig 10.4 shows a junction box with separate terminals for phase, neutral and protective conductors. The protective conductors in this case are contained within the sheaths of the wiring cables.

Fig. 10.5 shows a method for bonding the sheaths of metal-sheathed mineral-insulated cables. Fig. 10.6 shows the bonding lead necessary for compliance with IEE Regulation 543-02-07, when conduit systems are used to give earth continuity at socket-outlets. Fig. 10.7 shows the earthing terminal for a socket outlet for connection of the protective conductor. Fig. 10.8 shows a type of adaptor used on some portable appliances such as kettles and irons.

In certain locations, special hazards exist, for example where electric wiring or equipment is installed close to the metalwork of water or gas services, in bathrooms, kitchens and sculleries. In such locations there are often metal baths, sinks, tanks, radiators and exposed metal pipes in a damp

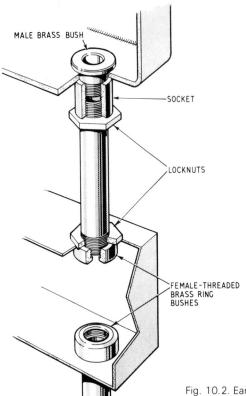

MALE BRASS BUSH

SOCKET

LOCKNUTS

FEMALE-THREADED
BRASS RING
BUSHES

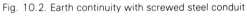

Fig. 10.2. Earth continuity with screwed steel conduit

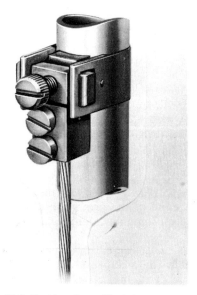

SAFETY EARTHING
EARTH
DO NOT REMOVE

Fig. 10.3. Earthing clamp (*Tenby*)

environment. If it is impracticable to segregate the electrical installation in such a way that there is no possibility of casual contact with such metalwork, then the metalwork must be bonded to the protective conductor. The object of bonding is to ensure that no dangerous voltage can exist on or between any two points on the exposed metalwork in the room.

Segregation may be provided by wood or other suitable insulating material when practicable.

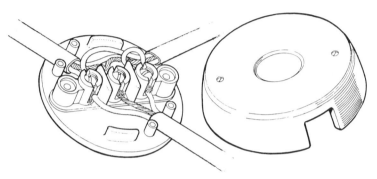

Fig. 10.4. Ring circuit junction box showing protective conductor connections (shaded wires)

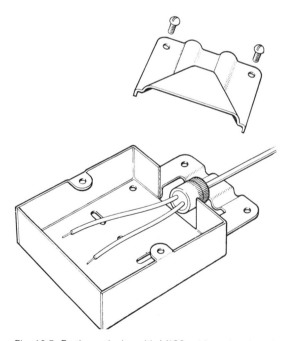

Fig. 10.5. Earth continuity with MICS cables, showing clamp fitting with switch box

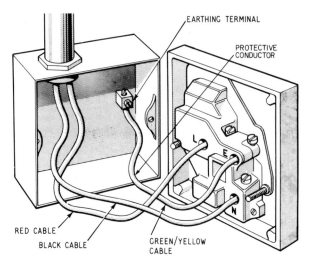

Fig. 10.6. Bonding wire used with conduit systems to provide staisfactory earth continuity at socket-outlets (IEE Reg. 543-02-07)

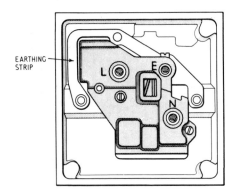

Fig. 10.7. 13 A socket-outlet showing earth (E) terminal connected to protective conductor and earthing strip to bond screws

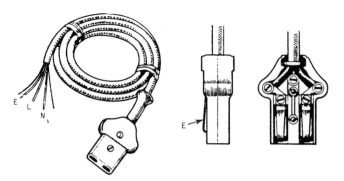

Fig. 10.8. Adaptor with external earth (E) contact which engages with metal frame on the appliance

Special precautions in bathrooms

IEE Regulations Section 601 specify additional precautions to be taken in rooms containing a fixed bath or shower. Some of these are:

1   All parts of a lampholder within 2.5 m of the bath or shower must be constructed of, or shrouded in, insulating material. Bayonet-type (B22) lampholders must be fitted with protective shields to BS 5042. Alternatively, totally enclosed lighting fittings may be used.

2   Every switch or other means of control or adjustment must be so situated as to be normally inaccessible to a person using a fixed bath or shower. This does not apply to electric shaver units complying with BS 3052, to insulating cords of cord-operated switches, nor to controls on instantaneous water heaters which comply with BS 3456, Section 3.9. No stationary appliance having heating elements which can be touched must be installed within reach of a person using the bath or shower. The sheath of a silica-glass-sheathed element is regarded as part of the element.

3   There must be no socket-outlets and no provision made for connection of portable apparatus (except electric shavers — see below).

4   Electric-shaver sockets complying with BS 3052 may be installed, the earth terminal being connected by a protective conductor to the consumer's main earth terminal. (Note: BS 3052 requires that the secondary circuit supplying the output sockets is isolated both from the supply mains and from earth.)

5   Circuits supplying equipment which is simultaneously accessible with other metalwork must have protective devices and earthing arrangements such that in the event of an earth fault, disconnection occurs within 0.4 seconds.

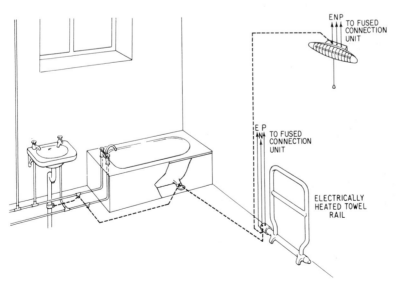

Fig. 10.9. Earthing in bathrooms. All exposed metal should be bonded to the protective conductor

Fig. 10.9 shows the bonding connection which should be made between the electric fire and towel rail earth terminals and other exposed metalwork, including the bath, hot and cold waterpipes and waste pipe.

Size of bonding conductor

The sizes for bonding conductors are specified in IEE Regulations Section 547.

Methods of protection from shock

One of the following basic systems may be used to protect users of electrical apparatus from shock or fire:

1   Direct or solid earthing of the apparatus.
2   Protective multiple earthing.
3   Omission of an earth entirely by using an all-insulated wiring system and electric appliances of all-insulated or double-insulated construction.
4   Employment of earth-leakage circuit-breakers.

Solid earthing

Earthing is direct or solid when the protective conductor is connected directly by a path of low impedance (such as a metallic conductor or the body of the earth) back to the neutral of the supply transformer. Protection is effected by the fault current causing operation of the fuse or overload circuit-breaker. There are two methods which may be employed for making a solid connection:

1   By connecting the earthing conductor to the metallic sheath or armour of the supply cable, or to a continuous earth lead in the supply cable or overhead line.
2   By connecting the earthing conductor to an earth electrode or electrodes, such as copper strip or rod, buried in ground. If this is the *only* means of earthing it is essential that all socket-outlets are protected by an earth-leakage circuit breaker (see pages 168–9).

Connecting the earthing lead to the metallic sheath or armour of the supply cable (with the consent of the supply authority) or to the earthing terminal provided, will usually provide a satisfactory earth, as the resistance will normally be low enough to pass the necessary fault current.

Connection of the earthing lead to water-pipes, either separately or jointly is not now permitted as the sole means of earthing an installation (IEE Reg. 542-02-04). This is because many modern water supply systems employ non-metallic mains, and renewal of old piping is often carried out with plastic piping and non-conducting joints. It should, therefore, be borne in mind that many existing properties that now employ a water-pipe for earthing may in the future require an alternative method.

In many rural areas where the supply is overhead, the only method of solid-earthing available is to connect to a metal electrode such as a driven rod or buried copper strip or plates.

Rod and strip electrodes Rod electrodes are made of either copper or wrought iron and are often provided with screw threads so that several lengths may be joined together in order to achieve a low earth impedance. Copper-strip electrodes should be buried in a straight trench as deep as possible. Rod electrodes are preferable as they can be driven deep into the soil, thus avoiding climatic changes of the top layers of the soil which may affect the earth resistance. The earthing lead can be connected to the electrode in a small pit, excavated for the purpose, or above ground. In both these cases, easy inspection is possible and there is little risk of corrosion. Where the earthing lead is connected to the earth electrode, a permanent label must be attached, indelibly marked with the words 'SAFETY ELECTRICAL CONNECTION — DO NOT REMOVE' in a legible type not less than 4.75 mm high — IEE Reg. 514-13-01 (Fig. 10.3).

The effectiveness of this type of electrode is dependent upon its resistance to earth. Obtaining a suitable low-resistance earth may prove too costly or impracticable owing to the high resistivity of the soil. In this event, protection can be achieved by using either earth-leakage circuit-breakers or protective multiple earthing (where the local electricity board adopts this system).

On domestic installations where the only means of earthing is by means of local earth electrodes (known as the TT system), i.e. where there is no metallic path back to the transformer neutral point, it is now compulsory for all socket-outlets to be protected by a residual-current-operated ELCB whose operating current multiplied by the earth fault loop impedance does not exceed the figure 50. (IEE Regulations 413-02-16.)

Protective multiple earthing (PME)

On this system the consumer's main earthing terminal is connected by the electricity board to the neutral conductor at the meter position on the consumer's premises. This means that a phase to earth fault becomes a phase to neutral fault (IEE Part 2).

The adoption of PME is at the discretion of the supply authority; the permission of the Secretary of State for Energy is required before it is introduced.

With PME working, as normally applied in the UK, the wiring on the premises to lighting points, socket-outlets etc. is normal i.e. the protective conductors and neutral are kept separate for all wiring on the consumer's side of the consumer unit. All exposed metalwork on the premises must be bonded to the main earthing terminal as described earlier in this chapter. This is most essential on PME working so as to prevent any difference in voltage between the metalwork of electrical appliances and any other extraneous metalwork should a break in the neutral occur outside the premises. Details of this bonding must be agreed with the electricity board, and the minimum conductor size is 6 mm^2.

The neutral conductor is usually connected by the electricity board to earth electrodes at a number of points along the route (hence the name PME), particularly when the supply is overhead, and this ensures that the neutral is earthed even if a break occurs along the route; this factor does not however directly concern the consumer.

The great advantage of PME is that a metallic path of very low impedance is provided back to the transformer neutral point, and this means that the consumer's overload protective devices will normally operate when a phase to earth fault occurs. Another big advantage is that this path is continuously monitored since a break in the neutral will be immediately noticed.

PME was originally used only in rural areas having overhead supplies where it was often difficult or impossible to obtain a low-impedance earth connection. It is however now being applied very widely, including urban areas where the supply is underground.

Special cables have been introduced for PME working and on some of these the neutral conductor and armour are combined.

Fig. 10.10 shows a typical consumer's installation with PME.

This system is known as the TN-C-S system as defined in Part 2 of the IEE Regulations.

In each consumer's installation, the protective conductors are brought back to an earthing terminal in the usual way, but the earthing lead (which must be insulated) is connected to the neutral-supply terminal by the electricity board. No fusible cut-out, automatic circuit-breaker, removable link or single-pole switch must be included in the neutral conductor (or any conductor which is connected to the neutral conductor) on the consumer's side of the supply terminals. This, of course, applies to all systems, whether PME or not, but is particularly important with PME working.

Protective conductors

Size of protective conductor
The methods of determining the size of a protective conductor are given in the IEE Regulations, Section 543.

In any event, if the protective conductor is separate, i.e. it does not form part of a cable and does not consist of conduit, ducting or trunking and is not contained within an enclosure formed by a wiring system, the conductor size must not be less than 2.5 mm² if mechanical protection is provided or is sheathed, or 4.0 mm² if there is no mechanical protection.

The IEE Regulations referred to above give two methods for determining the conductor size:

1   The size must be not less than the value obtained by the following formula (applicable where the disconnection time does not exceed 5 seconds)

$$S = \frac{\sqrt{I^2 t}}{k} \, \text{mm}^2$$

where S is the cross sectional area (square millimetres), I is the rms value (amperes) of the fault current, assuming a fault of negligible

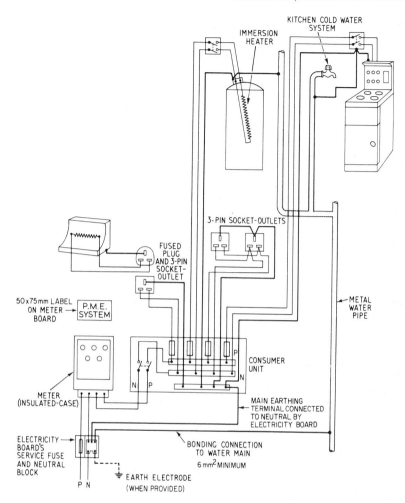

Fig. 10.10. PME system

impedance, which can flow through the protective device, t is the operating time in seconds corresponding to the value of I, and k is a factor depending on the material, insulation and other aspects of the conductor and on the initial and final temperatures. Values of k for differing types of protective conductor and various conditions are given in IEE Tables 54B to 54F.

If it is not desired to calculate the conductor size in accordance with the above method the size may be obtained by reference to Table 10.1 or IEE Table 54F. Where the protective conductor is of the same material as the associated phase conductors the size can be read direct from the table. Where the protective and phase conductors are not of the same material, the size of the protective conductor should be such that the required conductance, as read from the table, is obtained.

For both of the above methods, if a non-standard conductor size is obtained, the next highest standard size should be used.

Method 2 is the more simple to apply and will normally be satisfactory on domestic work. However, where PVC-insulated, PVC-sheathed twin and earth cables are used, it must be confirmed that the protective conductor is of the required size since in some cables it is of smaller cross-section than the other conductors.

In some cases method 1 may result in a smaller protective conductor size than method 2, particularly when the overload current protection consists of MCBs or cartridge fuses.

To take a typical example using method 2 we will assume a socket-outlet ring circuit fused at 30 A and using 2.5 mm² twin-and-earth PVC sheathed cable. From Table 10.1 the protective conductor will require to be 2.5 mm². If the sockets are wired on a radial circuit, using 4.0 mm² cable, the protective conductor will also require to be 4.0 mm².

Table 10.1. Minimum size of protective conductors in relation to size of associated phase conductors*

Size of phase conductor (S) (mm²)	Minimum size of protective conductor (Sp) (mm²)
Up to 16	S*
Above 16, up to 35	16
Above 35	S/2

* *Note:* Minimum size may apply—see IEE Reg. 543–1.
(*Based on IEE Table 54F*)

Types of protective conductor

The requirements for protective conductors are given in IEE Regulations Section 543-02. The items mainly affecting domestic installation are summarised below. For further details the Regulations should be consulted.

1   Where the protective conductor forms part of the same cable (flexible or non flexible) as the associated live conductors, it must comply with the BS for the cable.

2   The protective conductor of a ring final circuit (other than that formed by the metal enclosure or covering of a cable) must be in the form of a ring with both ends connected to the earth terminal at the origin of the circuit.

3   Where the protective conductor consists wholly or partly of the metal sheath and/or armour or other metallic covering forming part of a cable, it must be protected against damage or deterioration and must provide the required cross-sectional area in accordance with the relevant regulations. Flexible or pliable conduit must not be used as the protective conductor.

4   Where metal enclosures for cables are used as protective conductors they must comply with the following:

(i)   the cross-sectional area and any joints must meet the relevant

requirements
(ii) provision should be included for the connection of other protective
conductors as necessary

5   Where the protective conductor consists of the conduit, trunking,
ducting or the metal sheath and/or armour of cables, the earthing
terminal of each socket-outlet must be connected by a separate
conductor to an earthing terminal incorporated in the associated box
or other enclosure.

6   Protective conductors, not forming part of a cable or cable enclosure,
up to 6 mm² cross-section (excluding copper strip) must be insulated to
at least the equivalent of a single core non-sheathed cable of
appropriate size to BS 6004 and identified by the colours green and
yellow.

7   Connections of protective conductors must be accessible in accordance
with IEE Regulation 526-04-01 but this does not apply to joints in
conduit, ducting or trunking systems. Joints in conduit must be
mechanically and electrically continuous by screwing or mechanical
clamps. Plain slip or pin-grip conduits are not deemed to provide
satisfactory continuity.

8   Where the sheath of a composite cable is removed adjacent to joints
and terminations, the protective conductor must be insulated and
identified by the colours green and yellow.

Earth-leakage circuit-breakers (ELCBs)

Earth-leakage circuit-breakers are widely used where direct earthing is
impracticable, and where the PME system is not in use. Solid earthing is quite
useless if the total earthing resistance is so high that the fuses or circuit-
breakers will not operate when a short circuit to earth occurs. If the values of
earth-loop impedance required by the IEE Regulations cannot be obtained,
then an earth-leakage circuit-breaker must be used.

Also, as stated earlier in this chapter (page 164), IEE Reg. 413-02-16
requires that, on domestic installations where the only means of earthing is
by means of local earth electrodes, i.e. where there is no metallic path back
to the transformer neutral point, all socket-outlets must be protected by a
residual-current-operated ELCB. A similar ELCB is also required to
protect any socket-outlet intended to supply equipment outdoors, and such
sockets must be labelled. The operating current of the ELCB to protect
sockets used for outdoor equipment must not exceed 30 mA.

However residual-current operated ELCBs are now being much more
widely used, even in cases where they are not essential to meet the Regulations,
so as to provide extra protection. Their use is particularly beneficial in
locations such as garages, workshops, out-houses etc., where damp conditions
are likely and portable appliances may be used. They are available as separate
units or combined with socket-outlets (see Fig. 10.12) or built in to consumer
units — see Chapter 3.

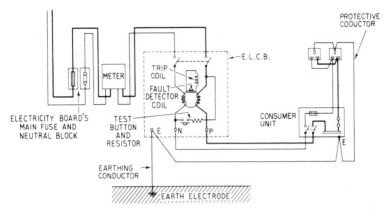

Fig. 10.11. Connection of residual-current-operated ELCB between mains supply and consumer unit

Fig. 10.11 illustrates the principles and use of residual-current-operated ELCBs. The device must be reset by hand even if the leakage current ceases to flow. The current rating of an ELCB must, of course, be equal to, or greater than, that of the fuse or overload circuit-breaker through which it is fed.

Residual-current-operated earth-leakage circuit-breakers
These are also known as a current-balance, core-balance or differential-current earth-leakage circuit-breakers.

The phase and neutral conductors act as two separate primary windings of a toroidal transformer core; the secondary winding of the transformer is connected to the trip coil (see Fig. 10.11). Normally, the currents in the phase and neutral windings are equal and opposite, so that no flux will be induced in the core and no current flows in the secondary winding or trip coil. On the occurrence of an earth fault, leakage current returns direct to the substation without passing through the neutral winding in the ELCB. The currents in phase and neutral therefore become unbalanced by an amount equal to the leakage current and the resulting secondary current energises the trip coil of the circuit-breaker. The circuit is opened when the earth-leakage current reaches the rated tripping current of the breaker. For ELCBs with low residual operating currents the transformer secondary voltage may be amplified before connection to the trip mechanism.

IEE Regulation 413-02-16 requires that when a residual-current device is used the product of the rated residual-operating current (in amperes) and the earth-fault loop impedance (in ohms) must not exceed 50.

When residual-current devices having low operating currents are being used, it may be necessary to consider the leakage currents which occur in normal use from equipment or cables, i.e. when no fault condition exists. Guidance on this aspect is given in IEE Regulation 531-02-03, and the total leakage in normal service on the installation protected by such a device must not exceed one half the nominal residual operating current of the device. This may affect the selection of the ELCB. However, this factor is not very likely to arise on domestic installations.

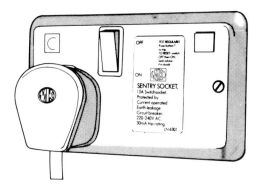

Fig. 10.12. 13 A socket-outlet with residual-current-operated ELCB (Note the test button in top right corner (*MK Electric Limited*)

It should be noted that a residual-current-operated ELCB will operate if a live terminal or conductor in the circuit concerned is touched by an earthed object or person, even if there is no contact from the live conductor to the earthed metalwork, provided the leakage current exceeds the operating current for the ELCB.

Test facility
A test button is included which introduces an artificial fault by switching in a resistance between the phase and neutral conductors. This feature enables a quick test to be made at any time to ascertain that the trip mechanism is working satisfactorily.

Note
Voltage-operated ELCBs are no longer permitted by the IEE Regulations and are therefore not covered in this book.

11 Inspection and testing

After completion of any new wiring installation or major alteration to an existing installation, the work should be inspected and tested to ascertain that there are no defects, and that all necesary items have been included.

However carefully an installation has been completed, it is always possible for faults to occur through nails being driven into cables, insulation being damaged, connections broken or incorrect, or even faulty apparatus being installed.

A full test of the completed installation should be carried out and the results carefully recorded. However, tests made during the progress of an installation at various stages can often detect faults which, if unnoticed, could prove costly to repair when the installation is completed. A quick test of all cables when they are delivered to site will ascertain whether they have any defects or have suffered damage in transit. When cables have been installed, a further test carried out before accessories are fitted will make certain that no damage has been done to the cables during installation.

The electricity board is entitled to refuse to connect a supply to a consumer if it is not satisfied that the installation complies with the relevant statutory regulations. In the UK an installation which complies with the IEE Regulations is deemed to meet the statutory requirements and should therefore be connected without difficulty.

If the board refuses to provide or maintain a supply to a consumer on the grounds that the installation is not considered satisfactory it must inform the consumer in writing of the reasons.

Inspection and tests to be carried out

The procedures for inspection and testing on completion of an installation are covered in the IEE Regulations, Part 7.

This gives a checklist for the visual inspection. In addition to the items listed, it should be checked that all the specified items such as socket outlets etc. have, in fact, been installed and that all equipment and plant are of the correct type, size and rating in accordance with the contract.

Following the inspection, tests should be carried out in accordance with the IEE Regulations quoted above. The tests normally applicable to domestic installations are listed below and it is important that these are carried out in the order shown so as to avoid danger from the test voltage should faults be present during the tests, e.g. a disconnection of the protective conductor. The IEE Regulations permit methods of testing other than those prescribed in the Regulations, provided they are equally effective:

(a) continuity of protective conductors and bonding
(b) continuity of ring circuits
(c) insulation resistance
(d) electrical separation and barriers
(e) polarity
(f) earth fault loop impedance
(g) earth electrode resistance
(h) operation of earth-leakage circuit-breakers

Continuity of protective conductors and bonding

The test voltage may be obtained from the mains supply via a transformer or from a hand generator or other portable device; this is often more convenient than a transformer since it permits the circuit wiring to be used for the tests as it can be disconnected from the supply.

The test voltage may be a.c. or d.c., but if d.c. is used it must be checked by inspection that no inductor is included in the protective conductor circuit since, on d.c., this would act purely as a resistance and the results would not be realistic.

Tests can readily be made by connecting the phase or neutral conductor of the circuit under test to the main earthing terminal, with the supply disconnected, and then testing between the phase or neutral and protective conductor terminals at each outlet on the circuit.

There are no particular resistance values for this test, which is only to check continuity. It may however be necessary to measure the resistance of the protective conductor in conjunction with the EFLI test — see later.

Continuity of ring final conductors

The requirement of this test is that each conductor of the ring is continuous. It is, however, not sufficient to simply connect an ohmmeter, a bell etc. to the ends of each conductor and obtain a reading or a sound.

So what is wrong with this procedure? A problem arises if an interconnection exists between sockets on the ring, and there is a break in the ring beyond that interconnection. From Figure 11.1 it will be seen that a simple resistance or bell test will indicate continuity via the interconnection. However, owing to the break, sockets 4 to 11 are supplied by the spur from socket 12 — not a healthy situation. So how can one test to identify interconnections?

The method is based on the measurement of resistance at any point across the diameter of a circular loop of conductor (Figure 11.2).

As long as the measurement is made across the diameter of the ring, all values will be the same. The loop of conductor is formed by crossing over and

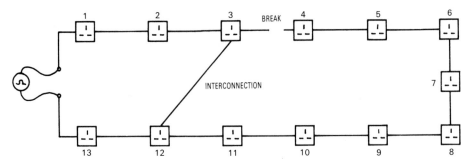

Fig. 11.1. Ring circuit with interconnection

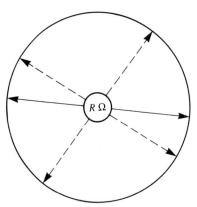

Fig. 11.2.

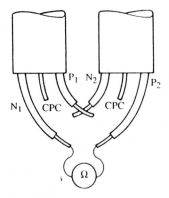

Fig. 11.3.

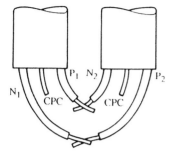

Fig. 11.4.

joining the ends of the ring circuit conductors at the fuse board. The test is conducted as follows:

1   Identify both 'legs' of the ring.
2   Join one *phase* and one *neutral* conductor of opposite legs of the ring.
3   Obtain a resistance reading between the other *phase* and *neutral* (Figure 11.3). (A record of this value is not important.)
4   Join these last two conductors (Figure 11.4).
5   Measure the resistance value between P and N at each socket on the ring. All values should be the same.

The basic principle of this method is that the resistance measured between any two points, equidistant around a closed loop of conductor, will be the same.

Such a loop is formed by the phase and neutral conductors of a ring dual circuit.

Insulation resistance

Insulation resistance tests must be carried out before the installation is permanently connected to the supply. Two tests are necessary (a) the resistance between all poles or phases and earth, and (b) the resistance between poles or phases.

Tests of insulation resistance are carried out by applying a d.c. test voltage approximately twice the normal r.m.s. value of working voltage; but it need not exceed 500 V d.c. for low-voltage circuits.

Testing instruments can be either the battery-operated type with a voltage-converter, or the hand-driven generator type. Both types of instrument have a scale in megohms ($M\Omega$) for insulation resistance testing, and usually a second scale in ohms (Ω) for continuity testing.

Large installations may be divided into groups of outlets, provided each group has not less than 50 outlets. For these tests, the term 'outlet' means every point and every switch, except that a socket-outlet, appliance or luminaire incorporating a switch can be regarded as one outlet. For example, a lighting point controlled by a separate switch constitutes two outlets, but a switched socket-outlet constitutes only one. Domestic installations can normally be tested as a single group.

Tests between conductors and earth

These tests should be made with all switches closed, all fuse links in position and all poles or phases of the wiring electrically connected together. At consumer units or switchfuses the fuses should be in position and, if possible, the main switch should be closed. The phase and neutral leads to the consumer unit from the supply side of the installation should then be connected together. Where it is not possible to test with the main switch closed (e.g. on installations with main switch already connected to the supply), the test may be made from the consumer's side of the main switch with the contacts open. In this case, the phase and neutral busbars should be linked together.

The insulation tester is then connected between the linked phase and neutral leads, or busbars, and the main earthing terminal. The insulation resistance to earth should not be less than 0.5 megohm ($M\Omega$). Where lighting circuits employ two-way switching, two tests should be made with one switch respectively in its alternative position; this ensures that both strapping wires are included in the test.

If the value obtained is less than $0.5\,M\Omega$, then the installation is not satisfactory and must be sectionalised until the faulty circuit is located. Before doing this all the fuses in the consumer unit should be removed and the outgoing neutral leads should be disconnected from the neutral busbar. The test should then be made again to ensure that the fault is not on the meter tails or busbars of the consumer unit. With the fuses removed and neutrals separated, each outgoing circuit should then be tested to earth, the phase and neutral conductors of each circuit being joined together in the consumer unit. Thus the faulty circuit can be located.

Earth fault tracing on looped circuits

Fig. 11.5 shows a method of splitting up a typical looped lighting circuit to trace the faulty section. With the lamps out and switches closed the complete phase and neutral conductors can be separately tested by connecting the test leads between earth and points T_1 and T_2 respectively. This will indicate

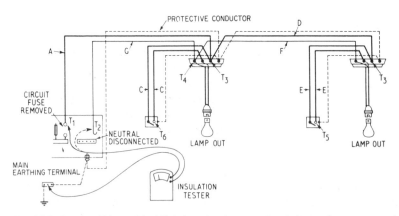

Fig. 11.5. Sectionalising typical lighting circuit to test insulation resistance to earth on individual conductors

which conductor has a faulty section. By breaking the connections at points T_3, and opening the switches, conductors A and D can be separately tested. Conductors F and G can be tested by breaking the connections at point T_4. By testing at points T_5 and T_6, with the switches open, the remaining switch wires C and E can be checked. Once the faulty section has been isolated and repairs or replacement carried out, the individual circuit and then the whole installation should again be tested.

Where separate off-peak circuits, including any floor-warming cables, are installed they should be linked with the other circuits for the above test.

Tests between conductors
This test should be made with the fuses fitted as for service, the main switch closed (where possible), all lamps removed, all current-using apparatus disconnected, and all local switches controlling lamps or apparatus closed. Where it is not practicable to disconnect any current-using apparatus or remove lamps, the local switches controlling them should be left open.

Where it is not possible to make the test with the main switch closed — e.g. where the tails between the main switch and the meter (or the earth-leakage circuit-breaker where fitted) are already connected and 'live' — then the main switch or switches should be left open and the test made from the consumer's side of the switch contacts.

The instrument is connected between the main phase and neutral tails which would normally be connected to the ELCB or meter; or, if this is not possible, between the phase and neutral busbars on the consumer's side of the main switch.

Where two-way switches are installed in lighting circuits, two readings should be taken — one with both switches in one 'on' position and another with both switches in the alternative 'on' position. Where a 2- or 3-phase supply is installed, tests should be made in turn from each conductor to all the other conductors connected together. Thus on a 3-phase supply four tests are required i.e.

(a) neutral to red, yellow and blue connected together
(b) red to neutral, yellow and blue
(c) yellow to neutral, red and blue
(d) blue to neutral, red and yellow

The insulation resistance for any test must not be less than $0.5\,M\Omega$.

Where current-using apparatus is disconnected for this test, an insulation test to earth should be made on it and the value should comply with the appropriate British Standard. If there is no standard applicable the insulation resistance should not be less than $0.5\,M\Omega$. Precautions should be taken against damage to any electronic equipment by the test voltage.

Tracing a fault between conductors
If the insulation resistance does not comply with the required standards it is necessary, first, to identify the faulty circuit, and then to locate and rectify the actual fault. The procedure is generally as described earlier for faults to earth. The fuses in the consumer unit should be removed and the neutral wires disconnected from the busbar and each circuit then tested separately

until the faulty circuit is found. Further tests should then be made on this circuit to locate the faulty section or equipment.

An aid to testers

When conducting insulation tests, a diagram of connections made at the commencement of the test is an aid to testers who are not continually engaged on testing such schemes. A note of each reading should be taken, for it should be remembered that the fault may not be concentrated on one particular lead, but may be spread over the system generally, due, for instance, to dampness which may cause leaks to occur at many points. Under such conditions, the only way to determine the case of the trouble is by careful comparison of all the readings taken.

An important precaution

After a section has been proved to be faulty and before an attempt is made to remove or to replace any of the wiring, it is advisable to examine and test all switches, ceiling roses, lampholders and connecting flexibles. Such is the excellence of present-day cables of reputable make that they are very unlikely to break down unless they have been damaged during installation. The fault is much more likely to be on the accessories such as switches, ceiling roses, etc.

The leads connecting the terminals of the testing instrument to the apparatus or system under test should be adequately insulated for the test voltage.

Electrical separation and barriers

These items are not likely to arise too often in domestic installations. One should confirm that IEE Regulations Part 7, as applicable, are met.

Check of polarity

The IEE Regulations require that (a) no fuse, or circuit-breaker other than a linked circuit-breaker, shall be connected in the neutral conductor and (b) single-pole switches must be connected in the phase conductor only, and any switch connected in the neutral conductor must also break the associated phase conductors. The reason for these requirements is, of course, that if a non-linked switch, fuse or circuit-breaker were connected in the neutral only, the element of an appliance or the lampholder in a luminaire would be live with respect to earth even if the fuse or circuit-breaker were removed or if the switch were open; this situation could be dangerous.

Also IEE Regulation requires that the outer contact of Edison screw or centre contact bayonet lampholders must be connected to the neutral conductor.

Polarity tests are required so as to confirm that the above conditions have been met. No specific methods are given in the Regulations for making these tests, but a number of simple methods are available.

On new installations the initial polarity tests must often be made with the system dead i.e. not connected to the mains supply. In fact, the supply will not normally be connected until tests have proved satisfactory. In such cases, a bell and battery set can be used for the tests as described below. The supply ends of the tails to the consumer unit should be left disconnected, all fuses fitted and all switches closed. Lamps should be removed and current-using appliances disconnected.

Bell and battery set
The simplest apparatus that can be used to carry out tests of polarity and continuity is the bell and battery set. A dry battery and bell are connected in series and the circuit completed by two test leads so that when the test leads are joined together the bell rings.

A quick test of conductor *continuity* can be carried out by connecting the bell set across the two leads left for connection to the meter, closing all lighting switches and socket switches and then shorting out the phase and neutral contacts at lighting points and socket-outlets. The bell should ring as each pair of contacts is shorted out.

Correct *polarity* can be determined by using one long fixed lead and one short test lead. The fixed lead is connected to the red phase lead which would normally be connected to the load side of the meter and the bell set with the short test lead is taken to each lighting point and socket-outlet in turn. The test lead is connected to the phase terminal and the bell should ring when the switch is closed and stop when the switch is opened. At unswitched socket-outlets, and at fuses or overload circuit-breakers, the bell should ring when the test lead is connected to the phase terminal.

The above polarity test does not prove that there is not a reversal between

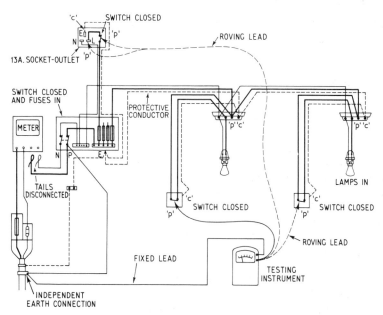

Fig. 11.6. Tests to prove polarity using one fixed lead and one roving lead from ohm-meter type instrument

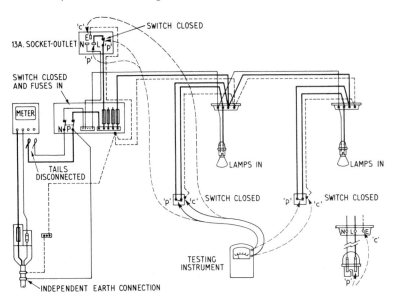

Fig. 11.7. Tests to prove either polarity or continuity with two roving leads from ohm-meter type instrument and utilising circuit wiring

the neutral and protective conductors. This is not very likely to occur but, if it does, the current which should flow in the neutral will, in fact, flow in the protective conductor and the circuit may well work normally. However this situation is undesirable and may be dangerous, particularly on socket-outlets, where the currents may be appreciable. Proof that the above reversal does not exist can be obtained by disconnecting the neutral of the circuit under test at the consumer unit and connecting one side of the tester to it. The other side of the tester is then connected in turn to each neutral point on the circuit. If the bell rings, this indicates that there is no neutral–earth reversal. The connections of flexible cords to plugs and appliances can readily be checked by simple tests between the plug and the appliance, using the bell and battery set.

A bell set indicates only correct connection and continuity — it cannot give indication of resistance values of protective conductors or of insulation. These must be determined by separate tests with a measuring instrument, as detailed elsewhere in this chapter.

Polarity and continuity tests can also be made by using proprietary testing instruments. These instruments usually have changeover switches which enable them to be used for other purposes such as testing insulation resistance and fault-finding.

Two main types of instruments are normally used: the hand-driven generator type, and the battery-operated type which has a transistorised converter. These are described later in this Chapter.

Figs. 11.6 and 11.7 show two methods of using a testing instrument to check polarity and continuity, including protective conductor resistance. For these tests, the instrument should be set to the 'ohms' scale. Fig. 11.6 shows a method using a long lead connected from the instrument back to the main earth, while a shorter test lead can be successively connected to various points as shown in the diagrams. This method is not always convenient, owing to the distance to be covered by the long lead connected to earth.

Fig. 11.7 shows an alternative method where the circuit wiring is utilised and two shorter test leads are connected to the instrument. Note that in both methods an independent test connection is made to earth. This is to ensure that the resistance of the main earth connection is included when these methods are used to check earth continuity.

To test polarity with the method shown in Fig. 11.6, the roving lead is connected successively to the points 'p', a resistance reading on the instrument indicating correct polarity; by connecting to points 'c', indication is given of protective conductor resistance. It will be noted that the resistance actually measured at points 'c' includes the resistance of the test leads; this should be measured on the instrument and deducted.

The method shown in Fig. 11.7 gives an indication of both polarity and continuity by connecting the two short test leads from the instrument to the points 'c' and 'p'. When testing at a switch, it must be closed and, when making tests at a ceiling rose, the switch controlling the point must be closed. Tests at two-pin sockets or ceiling roses are made by connecting 'p' to first one conductor and then the other, using the lower reading. The protective conductor resistance with this method of test includes the resistance of one conductor from the main switch to the point of test. Measurements slightly above or less than the specified value of protective-conductor resistance prove both continuity and correct polarity.

Polarity tests using mains supply
In some cases it may be desirable or convenient to carry out polarity tests with the supply connected to the installation. This situation can well arise when testing an existing installation or when making acceptance tests on behalf of the consumer on a new installation. A suitable filament type test lamp connected to two test leads may be used. A neon lamp is not suitable.

For testing polarity on lighting circuits, the lamps should be removed and one side of the test lamp connected to the neutral or protective conductor. The other lead from the test lamp should then be connected in turn to the switch terminals with the switch in the 'off' position. The lamp should glow when connected to one terminal of the switch but not when connected to the other. Tests at Edison screw or centre-contact bayonet lampholders should be made with lamps removed and switches 'on'.

Socket-outlets can readily be tested by means of a standard plug with a test lamp connected between the phase and earth terminals; this should be inserted into each socket outlet in turn. Glowing of the lamp confirms that the phase and neutral connections are not reversed. It does not however prove that there is not a reversal between the neutral and protective conductor connections although, as stated earlier, this is not very likely. A check on this can readily be made by temporarily disconnecting the neutral of the circuit under test at the consumer unit and then repeating the above tests at the socket-outlets. The test lamp should still glow; if it does not, a neutral-earth reversal is indicated. The neutral must be reconnected immediately after the test.

Checks that fuses, circuit-breakers etc. are connected in the phase conductor can readily be made by means of the test lamp at the consumer unit or other points as necessary.

Earth fault loop impedance (EFLI)

These tests are of the actual path taken by the current during an earth fault. This path is known as the 'earth fault loop' and the composition of this has been described in Chapter 10, Page 156.

Before the tests are made it is necessary to establish the maximum permissible values of EFLI for the outlets concerned. As indicated in Chapter 10 of this book, these can be obtained from the IEE Regulations. If the higher values of EFLI for socket-outlets permitted by IEE Regulations are being applied, the impedance of the *protective conductor* must be limited to the appropriate value in IEE, Table 41C in addition to the overall maximum value of EFLI; the impedance of the protective conductor between the outlet point concerned and the main earthing terminal must therefore be measured separately in these cases.

The EFLI tests should be made at all outlet points and if any results do not comply with the maximum values specified, remedial action must be taken.

The principle of the test is to pass a current from a known voltage source (usually the supply voltage) through a resistor of known value connected between the phase and protective conductors. The impedance is then obtained from the current value and can usually be read direct from a scale on the tester. A number of proprietary testers are available and certain of these can be plugged direct into the socket-outlet under test, using the normal mains supply to the socket. Test results should be carefully recorded and any variations should be investigated, even though the maximum permissible values are met.

It is important to check before these tests that the protective conductor circuit is satisfactory since, if it is disconnected, the full mains voltage may be applied to the protective conductor and thus to the metalwork of appliances.

The tests differ according to whether the wiring is in steel conduit or otherwise.

It should be noted that the IEE Regulations now refer to EFLI tests from

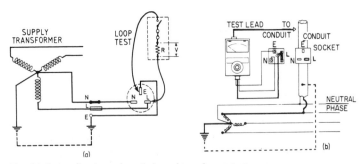

Fig. 11.8. Application of earth loop impedance test
(a) socket-outlet   (b) bonded metalwork

the phase conductor only (not the neutral) to the protective conductor. This is, of course, the actual path taken by the fault current.

A typical circuit for an EFLI test is shown in Fig. 11.8. The mains supply must be connected for this test. Tests at lighting outlets are similar to the test shown for bonded metalwork, the test lead being connected to the metalwork of the luminaire, when applicable, or other suitable point.

Earth electrode resistance

The test can be made either with current at power frequency supplied from the mains via a double-wound transformer, or with alternating current from an earth tester incorporating a hand-driven generator. If the test is made with current at power frequency, the resistance of the voltmeter used must be high — of the order of 200 ohms per volt.

Earth-testing instruments normally consist of a hand-driven generator or battery-operated device producing either a.c. or rapidly reversing d.c. Any stray currents present in the soil, e.g. due to leakage from distribution systems, will cause the pointer to waver at certain handle speeds. To obtain a steady reading, it is only necessary to increase or decrease the speed of turning.

The earth-electrode under test must be completely disconnected from all sources of supply other than those used for testing (see Fig. 11.9).

An alternating current of a steady value is passed between the electrode under test (T) and an auxiliary electrode (T_1). Electrode T_1 must be placed at a sufficient distance from T to ensure that the resistance areas of the two do not overlap — a distance of 30-50m is normally sufficient. A second auxiliary electrode T_2 is then driven into the ground half-way between T and T_1. Temporary earth connections can be made to 12 mm diameter solid mild-steel spikes driven into the soil to a depth of approx. 300-900 mm. The resistance of the earth electrode is then the voltage drop measured between T and T_2 divided by the current flowing between T and T_1.

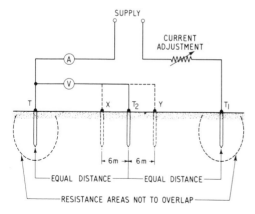

Fig. 11.9. Measurement of earth electrode resistance: T — earth electrode under test and disconnected, T_1 — auxiliary earth electrode, T_2 second auxiliary earth electrode, X — alternative position of T_2 for check measurement, and Y — further alternative position of T_2 for check measurement

To check that the resistance measured is a true value, two further readings should be taken with the second auxiliary electrode T_2 moved 6 m nearer to and then 6 m farther from T. If the three results are substantially constant, then the mean of the three readings can be taken as the resistance of the earth electrode T. If the results are not constant then the test should be repeated with T and T_1 farther apart.

Operation of residual current operated devices

These tests must be carried out after installation and connection, independently of the test facility included in the device.

In the recommended test for a residual-current-operated circuit breaker the test is made on the load side of the circuit-breaker, between the phase conductor of the circuit protected and the associated circuit protective conductor, so that a suitable residual current flows. All loads normally supplied through the circuit-breaker are disconnected during the test.

The rated tripping current shall cause the circuit-breaker to open within 0.2 seconds or at any delay time declared by the manufacturer of the device. (Note: Where the circuit breaker has a rated tripping current not exceeding 30 mA and has been installed to reduce the risk associated with direct contact as indicated in Regulation 412-06-02, a residual current of 150 mA should cause the circuit breaker to open within 40 ms.)

In no event should the test current be applied for a period exceeding one second.

The effectiveness of the test button or other test facility integral with the circuit-breaker is also to be tested, preferably after application of the externally applied tests described above. The IEE Regulations require that an ELCB should be tested at least quarterly by means of the test button.

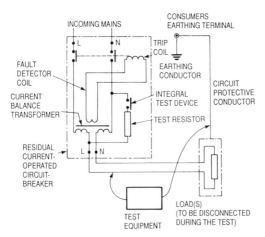

Fig. 11.10. Method of test for compliance with Regulation 613-16 for a residual current operated circuit breaker to BS 4293

Testers

Testers suitable for carrying out the foregoing tests are available from a number of manufacturers and some of these are briefly described below.

Fig. 11.11 shows two types of tester suitable for insulation and continuity tests.

The 'Wee megger' shown in (a) is a compact, self-contained, hand-operated type with a 500 V output, and is quite adequate for testing domestic installations. The test voltage is obtained from a hand-cranked, brushless a.c. generator, the output being rectified to give a d.c. voltage which is constant within a given range of cranking speeds.

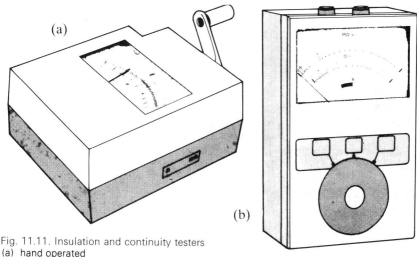

(a)

(b)

Fig. 11.11. Insulation and continuity testers
(a) hand operated
(b) battery operated (*Evershed and Vignoles*)

The tester in Fig. 11.11(b) operates on standard dry batteries, and the 500 V model is very suitable for testing domestic installations. It can be operated with one hand. The test voltage is generated electronically and is only available whilst the test button is pressed, thus conserving the battery supply. The selector switch has three positions i.e. battery check, megohms and ohms.

Fig. 11.12 shows a typical earth tester suitable for measuring the resistance of an earth electrode system as described earlier.

The tester shown is hand-cranked and the test voltage is a.c. The resistance values can be read direct from the scale in ohms. Battery-operated versions are also available.

Fig. 11.13 shows a tester suitable for measuring earth-fault loop impedance on domestic installations. This tester draws its power from the mains supply, and the values can be read directly in ohms. The instrument operates by passing a current of approximately 20 A for 30-50 seconds from the phase conductor through the EFLI path via a known resistor in the tester. The voltage drop across the resistor is measured in the tester from which the current and thence the EFLI is obtained. The result in ohms is read direct

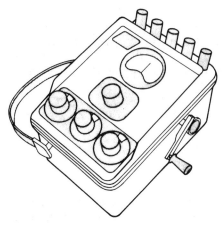

Fig.11.12. Hand-operated earth tester. A battery-operated version is also available (*Evershed and Vignoles*)

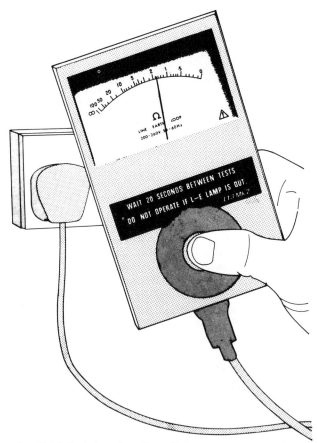

Fig. 11.13. Earth loop impedance tester (*Evershed and Vignoles*)

from the scale on the tester. The operation is not affected by variations in the supply voltage. The tester includes a recessed socket at the bottom for connection of the mains supply and a jack socket at the top for connecting the test lead when testing at lighting outlets, bonded metalwork etc. An automatic check of polarity and earth continuity is provided. In testing a socket outlet it is merely necessary to plug the tester, via the flexible cable, into the socket. In testing at other points the tester must be plugged into a socket outlet and then connected by a flexible lead to the point concerned. Fig. 11.8 shows the connections for two types of EFLI test.

Certification

On satisfactory completion of the inspection and tests the necessary certificates should be completed by duly authorised persons in accordance with Chapter 74 of the IEE Regulations. The Regulations also recommend that completion certificates include a recommendation that the installation be inspected and tested at stated intervals in the future.

Index

A.C. supply, 4–6
Adaptor, lampholder, 51
  socket-outlet, 71
Ambient temperature rating factor, 17

Bathrooms, precautions, 155
  bonding, 151–2
  heaters, 82–5
  lampholders, 50
  shaver units, 61
  socket-outlets prohibited, 61
  switches, 35, 45, 46
  water heaters, 105
Bell and battery set, 171
Bell circuit, 9
Bending spring, 132
Bending steel conduit, 129–130
  PVC conduit, 132
Bonding, 150
  bathrooms, 155
  hazard locations, 152
Box, ceiling, 48
  conduit, 121–5
  wall, 35–8
Burglar alarms see
  Intruder alarms
Busbar rising main, 6, 147
Bush, conduit, 123, 124

Cables:
  colours, 2, 72
  connection, 32
  current rating, 12–16
  floor-warming, 86–89
  for conduit, 119
  for lighting, 43
  heat-resisting, 43, 72, 105
  MICS see MICS cable
  PVC insulated, 15
  PVC insulated and sheathed, 15, 112, 135
  rising main, 6, 147

  sizing, 17–21
  under floorboards, 157–38
  voltage drop, 13–17
Ceiling rose, 47–9
Central heating system, 85
Circuit:
  bell, 9
  definition of, 1
  fluorescent tube, 52–9
  lighting, 1–2
  parallel, 1
  radial, 64
  ring, 8, 64–5
  three-wire, 1
Circuit-breakers:
  earth-leakage, 29, 161–63, 176
  excess-current, 3, 29–31
Circuit jointing, 41
Circuit looping, 39–40
Clock socket-outlet, 71
Colour coding, 2, 72
Conduit:
  cables for, 119
  capacity, 120
  non-metallic, 131
  under floorboards, 126
  watertight, 107–128
Conduit, PVC, 109–111
  bending, 132
  box connections, 111, 131
  capacity of, 120
  clips, 131
  coupling, 132
  earthing, 133–4
  fittings, 111
  fixing centres, 131
  flexible, 110, 133
  oval, 110
  saddles, 131
  temperature effect, 111
  threading, 109, 132

watertight joints, 133
wiring, 134
Conduit, steel:
bending, 129–130
boxes, 122–124
bush, 123
capacity of, 120
coupling, 121
earthing, 121, 152–53
elbow, 121
finishes, 107
installing, 119–33
joist recess, 126
oval, 107
plaster, in, 126
preparation, 127
saddles, 127
screwed fittings, 120
sizes of, 106
solid-drawn, 107
tee position, 121
thread cutting, 128
types, 106–7
watertight, 128
wiring, 130
Connection unit, fused, 66
Consumer units, 4, 29–31
Convector heater, 79
Cooker, wiring, 31, 73
Cooke's cutters, 131
Cord *see* Flexible cords
Current rating, 14–17

D.C. supply, 6
Detached building, isolation, 29
Dimmer switch, 45
Distribution boards, 3–4
Diversity factor, 13
Draw wire, 130

Earth-continuity methods, 151
conduit, 127, 151
MICS cable, 145
socket-outlets, 60, 158
Earth electrode system, 150, 161
Earth-fault-loop impedance:
definition, 149
maximum values, 150, 174
tester, 178
testing, 174
Earth-leakage circuit-breaker *see* ELCB
Earthing, 3, 148–63
PME, 6, 157
socket-outlets, 60
solid, 156
Edison screw lampholder, 51, 170
ELCB, current-operated, 29, 161
testing, 176
Electricaire system, 93–8
Electricity board service, 4–6, 28–9
Excess current protection:

by fuse, 2–3, 21–4, 29, 62, 64, 150
by MCB, 3, 24–7, 29, 150
Expansion couplers, 132

Fan Electricaire, 93
storage radiator, 89
Fan heaters, 80
Fire, fitting inset, 83
Fireplace conversion, 82
Fixed appliance, wiring, 73
Flats, supply to, 6
Flexible cords, 16
colour code, 72
heat-resisting 43, 46, 72, 88, 105
pendants, 46, 49
Floor-warming, 86
Fluorescent tube lighting, 52–9
circuits, 55–9
power factor capacitor, 55, 57
wiring, 58–9
Fused connection unit, 62, 66
Fuses, 2, 21–4
cartridge, 23, 24, 57
fluorescent lighting, 57
lighting circuit, 31, 40–1
main, 28
plug (13A), 61, 63
radial circuit, 64
renewal, 23
rewirable, 22
ring circuit, 60–3, 64–6
wire sizes, 24

Galvanised conduit, 107
Gland, MICS cable, 141
Gridswitch, 34–5
Grid system, 4, 5

Heaters:
bathroom, 85–6
fitting inset, 84–5
immersion, 100
size and position of, 81
thermostats for, 81
types of, 79
wiring, 82
Heating, underfloor, 86
Hilti tool, 86

Immersion heaters, 100
Indicator boards, 11
Inspection, 164
separation and barriers, 170
Insulating hot-water tank, 103
Intruder alarms, 9–11
Isolation of mains, 28–9

Joist recessing, 126
Junction box, 41–2
lighting circuit, 37–8
ring circuit, 67, 112

Lampholders, 49–52
Lighting circuits, 7–8
Lighting fittings, 45
Looping box, 122
Looping-in system, 39–41

Main fuse, 28
  switch, 28
Mains, rising, 6, 146
MCBs, 3, 24–7
Meters, consumers', 6
MICS cable, 113, 140
  clip spacing, 114
  current rating, 115–116
  end sealing, 140
  gland, 141
  installation, 140
  insulation test, 146
  pot fitting, 145
  roller straightener, 146
  stripping, 142
Motor:
  cable requirements, 27
  current, 14

Non-metallic conduit see
  Conduit, PVC

Off-peak electricity, 85
Ohm's law, 13
Oven thermostat, 77–79

Pattress block, ceiling, 45, 49
  wall, 36, 37
Plate switch, 34–5
Plug, 61, 63
  fuse, 3, 70, 71
Portable apparatus, 72
Power-factor capacitor, 54, 55, 57, 58
Protective conductors, 158
  definition, 149
  non-metallic conduit, 134
  size of, 160
  socket-outlets, 63
Protective multiple earthing, 6, 157
PVC cable, 112, 135
  connections, 136
  installation, 137
  multi-core, 15
  single-core, 15
  use in conduit, 106
PVC conduit see Conduit, PVC

Radial circuit, 64
Radiant heater, 79
  loading, 83
Radiator, oil-filled 80
  storage see Storage radiators
Rating:
  current, 13

factors, 16–17
  fuse, 17–19
Ratings, appliance, 12
  cable, 15–17
  motor, 13, 25
Ring circuit, 8, 64–8
  fuse, 64–6
  junction box, 64–4, 67–8
  spur, 66, 67, 84–85
  stationary appliances on, 66, 80, 82–84
Rising main, 6, 146
Room heating calculation, 83

Saddle spacing, 127, 131
Seal, MICS cable:
  fitting, 140
Service, electricity board's
  see Electricity board service
Shaver supply unit, 61, 84, 155
Sheathed wiring systems:
  all-insulated, 112, 135
  mineral insulated, 113, 140
Sheradised conduit, 107
Shower units, 105
Simmerstat control, 76
Socket-outlets:
  adaptors, 71
  British Standard, 62
  clock, 71
  combined with ELCB, 163
  connections, 62
  fitting, 68–70
  mounting height, 60
  proposed new standard, 72
  radial circuit, 64
  ring circuit, 64–7
  switch control, 71
  types, 61–3
  wiring, 66–72
  13A, 61, 62
Splitter unit, 29
Spurs, ring circuit, 66–7
Starter switch, glow-type, 53–4
  thermal type, 53–4
Steel conduit, see Conduit steel
Storage heater, water, 99
Storage radiators, 89
  Electricaire, 93
Supply, a.c., 4–6, 28
  d.c., 6
Switch box, 36–7, 124
Switch control, 7–8
  mounting height, 34
  position, 37
  wiring, 38
Switches:
  bathroom, 34
  branch, 32
  ceiling, 33, 34, 45–7
  dimmer, 45

Electricaire, 93
flame-proof, 34
flush, 35–7
heater-fan, 92
main, 28
surface, 38
three heat, 75
towel-rail, 84
watertight, 128
Switchfuses, 32
Switching, three-way, 8
two-way, 7, 43

Telephones, 72–3
Television aerials, 72–3
Testers, 177
Testing, 162
continuity of protective conductors and
bonding, 165
continuity of ring circuits, 165
earth electrode resistance, 175
EFLI, 174
ELCBs, 176
insulation resitance, 167
polarity, 170

Thermostat:
bimetallic tube, 78
capillary tube, 77
floor-warming, 86
heater, 81
immersion heater, 100
Three-phase a.c., 4–5
Three-way switching, 7, 8, 43
Towel-rail, 85
Two-way switching, 7, 43

Underfloor heating, 85

Voltage drop, 13–17

Wall-box, 35, 38
Water-heaters, 98
wiring, 105
Wire colour coding, 2, 72
Wiring fixed appliances, 73
Wiring systems:
all-insulated, 112, 135
choice of, 117
MICS, 113, 140
PVC conduit, 109–131
steel conduit, 106, 118